# Saccharomyces

*Edited by Thalita Peixoto Basso and Luiz Carlos Basso*

Published in London, United Kingdom

IntechOpen

*Supporting open minds since 2005*

Saccharomyces
http://dx.doi.org/10.5772/intechopen.90981
Edited by Thalita Peixoto Basso and Luiz Carlos Basso

Contributors
Muhammad Jahangeer, Zahed Mahmood, Areej Riasat, Mehvish Ashiq, Naveed Munir, Shakira Ghazanfar, Amanda Sena, Gustavo Monteiro, Maria Araújo, Paula Barbosa, Marcelo Mello, Tonny Leite, Sandra Assis, Michael Thomas Fasullo, Nilay Seyidoglu, Cenk Aydin, Seyma Hande Tekarslan-Sahin, Yanni Sudiyani, Eka Triwahyuni, Muryanto, Roni Maryana, Muhammad Eka Prastya, Muhammad Numan, Muhammad Asad, Usman Ali, Mahwish Salman



First published in London, United Kingdom, 2021 by IntechOpen
IntechOpen is the global imprint of INTECHOPEN LIMITED, registered in England and Wales, registration number: 11086078, 5 Princes Gate Court, London, SW7 2QJ, United Kingdom
Printed in Croatia

British Library Cataloguing-in-Publication Data
A catalogue record for this book is available from the British Library

Additional hard and PDF copies can be obtained from orders@intechopen.com

Saccharomyces
Edited by Thalita Peixoto Basso and Luiz Carlos Basso
p. cm.
Print ISBN 978-1-83968-789-1
Online ISBN 978-1-83968-790-7
eBook (PDF) ISBN 978-1-83968-791-4

For EU product safety concerns:
IN TECH d.o.o., Prolaz Marije Krucifikse Kozulić 3, 51000 Rijeka, Croatia, info@intechopen.com or visit our website at intechopen.com.

# Meet the editors

Thalita Peixoto Basso received her bachelor's degree in Agriculture Engineering. During this period, she studied the fermentation characteristics of *Saccharomyces cerevisiae* isolated from ethanol industrial processes. She obtained her master's degree from the University of Sao Paulo (ESALQ/USP). During this time, she isolated and selected fungi with high cellulose activity for enzymatic hydrolysis of sugarcane bagasse. Dr. Basso received her Ph.D. in the Agricultural Microbiology Program at ESALQ/USP, with a period as a visiting scholar at the University of California Berkeley and Energy Bioscience Institute. Meanwhile, she worked on the improvement of *S. cerevisiae* by hybridization for increased tolerance toward inhibitors from second-generation ethanol substrates. Currently, she is a collaborating professor and a postdoc working with metabolomics and proteomics of fermentation processes at the Genetics Department of ESALQ/USP.

Luiz Carlos Basso obtained a bachelor's degree in Agriculture Engineering and a master's degree in Soil and Plant Nutrition from the University of Sao Paulo (ESALQ/USP), in 1969 and 1973, respectively. He obtained a Ph.D. in Biological Science from São Paulo State University (UNESP), Brazil, and postdoctoral degrees from Institut des Produits de la Vigne, Montpellier (1989), and Superior Technical Institute, Lisbon. Since 1980, Dr. Basso has been involved with yeast biochemistry and physiology, aiming to increase ethanol yield using the fed-batch industrial process. During the last fifteen years, he has conducted a yeast selection program resulting in the most widely used *Saccharomyces cerevisiae* strains (PE-2 and CAT-1) in the Brazilian ethanol industry. Currently, Dr. Basso is a senior professor at ESALQ/USP teaching Biochemistry (for undergraduate students) and Biochemistry and Physiology of Yeast Fermentation (for graduate students) and selecting tolerant strains for lignocellulosic inhibitors.

# Contents

# Preface

The application of microorganisms to produce novel bioproducts, as well as bioproducts with economic importance already established, has great value for the economy and the environment in achieving sustainable production worldwide. The *Saccharomyces* genus has considerable importance in the production of several biomolecules by fermentation with vast applications in industries such as agriculture and pharmaceuticals. Additionally, *S. cerevisiae* is a model system for studies of anticancer agents and has great relevance in medical research, being crucial to understanding tumor cells mechanisms.

This book, *Saccharomyces*, provides a comprehensive overview of the current state of the art of the use of the *Saccharomyces* genus in important areas such as agriculture and pharmaceuticals. This book reflects the value of this microorganism in related areas. It includes seven chapters in two sections: "Agricultural and Biotechnological Applications" and "Medical and Pharmaceutical Applications." Chapters in the first section discuss metabolic engineering of *S. cerevisiae* using CRIPR-Cas9 technology for producing biopharmaceuticals, fruit juice fermentation for antioxidant activity, and the mode of action of indigenous *S. cerevisiae*. Chapters in the second section discuss the performance of *Saccharomyces* as an antiviral experimental microorganism on pandemic diseases, application of yeast to study DNA repair and damage tolerance on cell cycle division, how calorie restriction can support the anti-aging process using yeast budding cells, and secondary metabolites from *S. cerevisiae* with anticancer activity.

I hope this book will be interesting for researchers around the world interested in the *Saccharomyces* genus. I also hope these microorganisms can be applied to diverse bioprocesses to produce novel bioproducts useful for new discoveries.

**Thalita Peixoto Basso and Luiz Carlos Basso**
University of São Paulo,
Brazil

Section 1

# Agricultural and Biotechnological Applications

Chapter 1

# Metabolic Engineering of *Saccharomyces cerevisiae* for Industrial Biotechnology

*Seyma Hande Tekarslan-Sahin*

## Abstract

*Saccharomyces cerevisiae* is an important and popular host for production of value-added molecules such as pharmaceutical ingredients, therapeutic proteins, chemicals, biofuels and enzymes. *S. cerevisiae,* the baker's yeast, is the most used yeast model as there is an abundance of knowledge on its genetics, physiology and biochemistry, and also it has numerous applications in genetic engineering and fermentation technologies. There has been an increasing interest in developing and improving yeast strains for industrial biotechnology. Metabolic engineering is a tool to develop industrial strains by manipulating yeast metabolism to enhance the production of value-added molecules. This chapter reviews the metabolic engineering strategies for developing industrial yeast strains for biotechnological applications and highlights recent advances in this field such as the use of CRISPR/Cas9.

**Keywords:** metabolic engineering, evolutionary engineering, *Saccharomyces cerevisiae*, biopharmaceuticals, industrial biotechnology, CRISPR-Cas9

## 1. Introduction

The term "metabolic engineering" was introduced by [1] into the science. Metabolic engineering is defined as "the improvement of cellular activities by manipulation of enzymatic, transport and regulatory function of the cell with the use of recombinant DNA technology" [1].

Metabolic engineering aims to manipulate genetic information of the strain and "improve the cellular activities" of the strain [1]. Stress tolerance of yeast is important and needs to be improved in industrial processes. Heterologous pathways and metabolic engineering cause stress on the yeast strain. The compound of interest can be toxic to the host strain and heterologous pathways are more sensitive than endogenous pathways. During the industrial processes, high salt, high temperature, high ethanol, acids and inhibitors can cause stress and affect industrial processes [2].

Metabolic engineering makes stress tolerance possible by improvement and modification of cellular functions of yeast [3]. Metabolic engineering can be subdivided as rational metabolic engineering, inverse metabolic engineering and evolutionary engineering.

## 1.1 Rational metabolic engineering

Rational metabolic engineering is the fundamental type of metabolic engineering. It focuses on the engineering of proteins and enzymes based on the knowledge of pathways and their regulation [4]. Protein activities are optimized to design a desired strain based on the protein and host information [5].

Application of systems biology helps to obtain protein and host information and also to model the system. In rational metabolic engineering, a mathematical model is needed to predict the strategies that can improve the strain [6].

## 1.2 Evolutionary engineering

Evolutionary engineering is the method for strain improvement by mutagenesis or gene recombination and shuffling, after which a cell with the desired phenotype can be obtained. In other words, multiple cycles of random genetic perturbation are performed and the strains are selected. These two events are sequentially performed [7, 8]. Evolutionary engineering is a method that improves the strain by mimicking the evolutionary process [9, 10]. Aim of the evolutionary engineering is to obtain desired phenotypes by mimicking the natural evolutionary process. The evolutionary process is achieved by appropriate selective pressure. Their molecular mechanism is then studied. Industrially important traits such as stress tolerance, product formation and substrate utilization are improved by evolutionary engineering. Through evolutionary engineering, *Saccharomyces cerevisiae* can be made to become resistant to multiple types of stress. Evolutionary engineering is used to improve the stress resistance of *S.cerevisiae*, such as ethanol resistance [11], salt resistance [12], freeze resistance [13].

In this method, the best screening method must be found for the stress of interest. Genetic modifications are random. Strains evolved through evolutionary engineering can provide genetic information for the improved strain and this information can be used for inverse metabolic engineering [7, 14] or random methods for deletion or overexpression of genes [15] or to introduce random mutations. It accumulates random mutations in the genome of the host. It is not easy to determine which genetic modification causes the strain improvement [16].

While developing a microbial strain, toxicity and tolerance of bioproduct, cell growth during fermentation and also downstream processes are considered. Optimization of these factors and cell performance is a difficult task due to the lack of knowledge on the relationship between genotype and phenotype.

Evolutionary engineering can be classified into two categories: adaptive laboratory evolution (ALE) and directed evolution [17].

### *1.2.1 Adaptive laboratory evolution*

The physical and chemical mutagens and error prone-PCR (ep-PCR) are the methods of traditional random mutagenesis used in adaptive laboratory evolution.

In adaptive laboratory evolution, after strains are cultured under selected pressure for long periods, strains with beneficial mutations are screened and selected. Cells with beneficial mutations outgrow the other strains during culturing under the selection pressure. Adaptive laboratory evolution (ALE) is commonly used to improve product titer, carbon source utilization, increase tolerance, and optimize cell growth and identify unknown biological mechanisms [17, 18].

### *1.2.2 Directed evolution*

In directed evolution, the desired selection pressure is applied to develop enzymes with new or improved properties. Protein engineering uses directed evolution to enhance activity of the enzymes. Directed evolution focuses on a gene encoding a protein or enzymes. ALE focuses on the entire organism and exhibits spontaneous mutations. Genetic diversity can be generated at a single gene, at pathways or for the whole genome [17, 19].

Oligo mediated targeted mutation generation such as multiplex automated genome engineering (MAGE), a modified method of MAGE called yeast oligo-mediated genome engineering (YOGE) and RNAi-assisted genome evolution (RAGE) can be used for directed evolution [20, 21].

## 1.3 Inverse metabolic engineering

Inverse metabolic engineering has three important steps to be applied to the strains. In the first step, the desired phenotype is identified, constructed and calculated. In the second step, the phenotype of interest is characterized according to genetic or environmental factors. In the third step, the phenotype of another strain or organism is translated to our strain by directed genetic manipulations or environmental manipulations [4]. Inverse metabolic engineering benefits from phenotypic differences. The host strain is exposed to different environmental conditions. Then, the trait of the organism which makes it resistant is investigated; after which, genetic basis of this trait is identified. Transcriptomics, proteomics and metabolomics is used to identify the basis of the trait [14, 22]. Xylose assimilation is improved in recombinant *Saccharomyces cerevisiae* by inverse metabolic engineering. A genomic fragment library from *Pichia stipitis* was introduced into *S.cerevisiae* expressing XYL1 and XYL2. Then, the transformants with high xylose growth rate were chosen. After sequencing, XYL3 is the responsible gene for high xylose growth rate [23]. The concept of reverse metabolic engineering has advantages. There is no information about the proteins and their regulation in the pathway. Their regulation, industrial strains and actual production conditions can be directly utilized to identify key genetic players. Homologous genes are responsible for the final strain development. New genetic targets can be discovered. When heterologous genes are needed to be added, it is better to use rational metabolic engineering. Despite the advantages of inverse metabolic engineering, strain development will be more successful when all metabolic engineering methods are used [9].

In theory, microorganisms can produce all the metabolites that they produce in their cells, however they produce low levels of these products. Metabolic engineering has made it possible to induce the production of chemicals and proteins in higher volume. Titer, yield and productivity of the products is improved by metabolic engineering. The improvement of results for the product of interest contributed to the development of the strategies in metabolic engineering. To develop and improve industrial strains, the strategies with the help of metabolic engineering, systems biology and synthetic biology are used [24–27].

There are metabolic engineering tools that help us to make industrial strains. Using metabolic engineering, novel metabolic pathways can be constructed, new metabolic engineering targets can be identified, gene expression can be controlled and tolerance to stress can be increased. These tools and strategies can be used not only for strain development but it will be helpful for fermentation strategies, as well. Systems biology, synthetic biology and metabolic engineering help the industry develop resistant and efficient strains [27].

## 2. Systems biology and metabolic engineering

Systems biology focuses on interpreting cellular networks via computational simulations and omics data analysis [28].

Metabolic engineering research whereby an engineered strain produces chemicals, proteins, biofuel or a material that has economical value. It is important to scale up the strains for industry. Metabolic engineering is successful in developing a strain that overproduces the product of interest at the lab-scale. However, developing an industrial strain can produce bioproducts is challenging and takes time, effort and money. A combination of metabolic engineering and systems biology is systems metabolic engineering [24, 29].

Traditional metabolic engineering approaches are integrated into systems biology and synthetic biology. Systems biology focuses on genome-scale computational simulation and omics analysis and synthetic biology focuses on the tools at the molecular level and pathways. Genome engineering and evolutionary engineering focuses on stress tolerance [28].

The combination of metabolic engineering and systems biology is called system metabolic engineering. It focuses on cell growth and target chemical production to accomplish the desired phenotype. Omics technology can be used for engineered strains. Sequencing all genomes is helpful to understand the differences between strains, however the difference is large. It will be difficult to find the real reason for the phenotype and the cost will also be high. The analysis of transcriptome, proteome, metabolome and fluxome offer information about the differences between strains because they are closely related to the phenotype [9]. The last step of inverse metabolic engineering is "omics" technologies. Here, "omics" technologies can find the differences at the gene level. The control of the target genes can be done by deleting or overexpressing the target genes in the strain. The result of this experiment will determine if this genetic modification will give the desired phenotype. The aim of the omics technologies are the gene detection (genomics), mRNA detection (transcriptomics), proteins detection (proteomics) and metabolites detection (metabolomics) in a strain. Omics technologies are genomics (study of an organism's genome, detection of genes), transcriptomics (mRNA detection, gene expression microarrays), proteomics (detection of protein to understand pathways and networks) and metabolomics (detection of global metabolite profiles in a system) and fluxomics (detection of metabolic fluxes) [30]. Genomics is the study of the whole genome of an organism. The study of chemical process, metabolites and the products of metabolism in a cell is called metabolomics. Analyses of omics can provide information on cellular and metabolic characteristics at the industrial strains. Omics data can give information on which genes or pathways are enhanced for the production of the bioproduct of interest [27]. Multi-omics data is applied to investigate the characteristics of the strain. Multi-omics analyses are done to select genetically engineered strains and the best strain for metabolic engineering.

Genome-scale metabolic models (GEMs) simulate the gene-protein-reaction (GPR) relationships for all the genes in an organism and help the researchers to predict metabolic response and fluxes for various systems-level metabolic studies. Genome-scale metabolic model (GEM) helps to develop strains to produce chemicals and drugs. GEM helps to predict enzyme functions, interactions among cells or organisms and understand the human diseases [31].

After the whole genome sequencing, genome-scale metabolic models have helped to predict cellular metabolisms and function and showed a way to identify the targets to increase the compound of interest. The first GEM for *Haemophilus influenzae* was established in 1999. MOMA is one of the widely used algorithms to identify the targets which helped to increase the production of the compound of interest. OptKnock is another program used for metabolic engineering [27].

The synthetic pathways for the compound of interest can be predicted by pathway prediction algorithms. Biochemical network integrated computational explorer (BNICE), RetroPath, GEM-Path, OptStrain and DESHARKY are the examples of pathway prediction algorithms [27].

Stable or enhanced enzymes are needed for the processes of the production of the chemicals. Computational protein design tools design new or improved enzymes by identifying core parts of protein structure and target sites for engineering [27].

## 3. Synthetic biology and metabolic engineering

Synthetic biology is the combination of engineering and biology. It can redesign or engineer biological systems [32]. In synthetic biology, biomolecular components, networks and pathways are designed and then, used to reprogram organisms [33]. Synthetic biology focuses on DNA synthesis, design and construction of novel metabolic pathways [27]. After construction of the synthetic pathways, the metabolic pathways should be optimized and maximized to improve the yield of the product of interest. The target site of metabolic pathways was chosen by the help of genome-scale metabolic models (GEMs) and multi-omics analyses. The high performing, resistant and efficient strains will be constructed according to synthetic biology tools [17, 27]. Optimization of the pathway needs to done after the synthetic pathway is constructed. Pathway optimization can be controlled by gene expression. Gene expression can be modulated by gene expression components or regulatory RNAs. Any modulation on the 3′ or 5′ untranslated region (UTR), transcription factor, promoter, ribosomal binding site or terminator can control gene expression. RNAi will be effective in controlling genes [34]. RNAi is a gene knockdown system in eukaryotes and an important method for metabolic engineering. RNA-induced silencing complex (RISC) protein reduces mRNA levels via a small interfering RNA (siRNA). Double stranded RNA is degraded to siRNA by a protein called dicer. Argonaute recognizes small guide RNAs which will recognize and degrade mRNAs of the target gene. RNAi has been widely used for metabolic engineering in eukaryotic organisms. In prior study [34], hairpin RNA expression cassettes are constructed to improve itaconic acid production.

The disadvantage of metabolic engineering is whole-genome sequencing to identify the trait. Clustered regularly interspaced short palindromic repeats (CRISPR-Cas9) has led to a new era for genome engineering [35]. CRISPR-Cas9 made a huge impact on the advancement of engineering of microbial cell factories. Endogenous homology-directed repair (HDR) or non- homologous end joining (NHEJ) are DNA repair pathways and used to insert or delete genes. However, their efficiency is low for metabolic engineering of microbial organism [36, 37, 38].

Endonucleases are used to double strand break induction and increase the recombination efficiencies [39]. The disadvantage of endonucleases is that they are not a successful strategy for large genomes. Transcription-activator nucleases TALENs and Zinc-finger nucleases (ZFNs) are advantageous for sequence specificity [38]. TALENs are focused on a smaller size of the genome. The CRISPR-Cas9 system is a defense mechanism that bacteria against viruses. In a class II CRISPR/Cas system, Cas9 is an endonuclease and introduces a double strand break. Trans-activating CRISPR RNA (tracrRNA) and CRISPR RNA (crRNA) collectively guide Cas9 to the target region. Cas9 cleaves the DNA-strand which is the complementary strand of the crRNA-guide sequence. In genome engineering, crRNA and tracrRNA are fused and used together.

The CRISPR-Cas system is used for metabolic engineering in *S. cerevisiae* [40]. CRISPR interference (CRISPRi) is a system that is used for gene downregulation. Deactivated Cas9 binds to the target DNA and blocks the transcriptional initiation [41].

## 4. Biopharmaceuticals and metabolic engineering

Bio-based chemicals' market size is estimated to reach USD 97.2 billion by 2023. According to biopharmaceutical market forecast report, the market size is 239.8 billion in 2019 and it is estimated to grow at a CAGR of 13.28% during period 2020–2025. Biopharmaceuticals represent 25% of commercial drugs and about 40% of total pharmaceutical sales [42]. Total biopharmaceutical sales are over $275 billion in 2019 and doubled from $125 billion in 2012. Total biopharmaceutical sales are growing at 12 percent annually. $200 billion of $275 billion are the sales of recombinant proteins. The rest of the sales are non-recombinant vaccines as well as blood and plasma products [43]. Sales of most biopharmaceuticals have grown significantly versus sales in 2011. The top three categories of sales are recombinant proteins, monoclonal antibodies and insulin [44]. The USA FDA and European Medicines Agency have stimulated the biopharmaceutical industry to produce more biopharmaceuticals [43]. Improvement of human health and life longevity are the benefits of the medicine. Biopharmaceuticals play an important role in the treatment of many disease [45]. The first biopharmaceutical was the recombinant human insulin produced in *Escherichia coli*. Eli Lilly launched the recombinant human insulin to the market in 1982 [46].

Approximately 70% of potential drugs in development are developed for diseases such as diabetes, cancer, neurological and immunological diseases. The effectiveness of biopharmaceuticals in the treatment of cancer and HIV/AIDS has been observed in the last decade. Deaths due to these diseases have decreased with the use of biopharmaceuticals. As a result, it has led to an increase in the use of biopharmaceuticals in the global market [47].

Therapeutic enzymes, therapeutic proteins, recombinant growth factors, cell and gene therapies, recombinant hormone, synthetic immunomodulators, hormones, monoclonal antibodies and vaccines are biopharmaceuticals, and they have been extensively used as therapeutic agents [45]. The term "biopharmaceuticals" was named in the 1980s. It means pharmaceuticals that are produced using genetic engineering [48]. Biopharmaceuticals have many advantages such as targeting only specific molecules, having fewer side effects, having high specificity and high activity [49]. Biopharmaceuticals are 100–1000 times larger than conventional drugs. It is necessary to use microbial and mammalian cells for the production of all therapeutic proteins. As a result, biopharmaceuticals are produced in *E.coli*, yeast and mammalian cells [50].

Yeast has advantages among the cell factories that are used for biopharmaceutical production. Yeast needs inexpensive medium compared to mammalian cell culture. Inexpensive medium for yeast reduces the cost of the biopharmaceutical production. Fermentation technologies that are used for yeasts are well-known and well-established [51]. However, yeast also has disadvantages in the production of therapeutic proteins. The high glycosylation capability of yeast is a disadvantage. If the high glycosylation capability is blocked, yeast can produce therapeutic proteins with the humanized glycosylation [52].

Cell factories are advantageous due to product quality, scale-up and downstream processes. The advantages of producing proteins in mammalian cells are having properly-folded protein, good pharmacokinetics and human-like N-glycosylation. However, mammalian cells are sensitive to bioprocessing. and the growth medium of mammalian cells are expensive [53]. The first human insulin is produced in *E.coli*. Protein production in *E.coli* is advantageous due the use of and inexpensive medium, fast growth time, high cell density culture, easy transformation and fast protein production rate. The disadvantage of protein production in *E.coli* is incorrect folding, low solubility and secretion of the protein. *S.cerevisiae* has good properties such as proper folding, easy culture growth and correct post-translational modifications. Secretion of the product to an extracellular medium is possible and this makes the protein purification easier. *S. cerevisiae* is also free of pathogens. *S.cerevisiae* is the most commonly used yeast strain for recombinant protein production. *S.cerevisiae* can be defined as predominantly unicellular however, many yeasts have both unicellular and multicellular lifestyles [54]. *S.cerevisiae* is used as a model for diseases. *S.cerevisiae* was the first eukaryotic organism to be fully genome sequenced [55]. *S.cerevisiae* has tolerance to chemical and physical stresses and that makes it a good model organism protein production in the industry [54].

Bioproduct process contains four stages: (1) first upstream process, (2) second upstream process, (3) midstream process and (4) downstream process. Fermentable carbohydrates are converted in the first upstream processes. A high-performance strain is developed in the second upstream process. In the midstream process, the strain is grown, and it produces the product of interest. The desired product is purified in the downstream process. Microorganisms should be optimized to produce the chemical or material of interest efficiently. The optimization and modifications can be applied on the microorganisms by metabolic engineering [17].

Metabolic engineering uses synthetic biology, systems biology, and evolutionary engineering to develop microorganisms. Metabolic engineering develops highly efficient microorganism strains to produce important products such as bioproducts, bulk chemicals, materials, natural products, fine chemicals, and polymers. Target products should be selected first. Biofuels, bulk chemicals, polymers, fine chemicals (including drugs), materials and natural products are bioproducts. Bulk chemicals are chemicals with low price and produced in large quantities. Fine chemicals are more expensive than bulk chemicals and are produced in small quantities.

Traditional petroleum-based plastics are produced from fossil fuels and are unsustainable; thus, bioplastics have become popular [56]. Bio-based polymers are polymers that can replace petroleum-based plastics. Polyhydroxy-acids (PHA) are biodegradable natural polyesters that are produced in microorganisms and have biodegradability and biocompatibility properties. To produce more PHAs and poly (lactic-co-glycolic acid), microorganisms can be engineered by metabolic engineering. Biofuels are produced by biological and chemical processes. Fatty acid biosynthetic pathways, ethanol pathways, CoA-dependent reverse b-oxidation pathways, keto acid pathways and isoprenoid pathways are key pathways for biofuels. Bioethanol is one such biofuel that is produced by microorganisms and developed by metabolic engineering. *E. coli* and *Saccharomyces cerevisiae* have been used for the production of biofuels. The classification of natural products is based on their structures. This classification contains alkaloids, terpenoids, phenylpropanoids, and polyketides. The source of natural products are natural sources. The cost of extracting natural products is high. The extraction of natural products leads to low yield. Natural products can be produced by chemical synthetic routes. However, production by chemical synthetic routes can generate multistep reactions and stereoisomers. Natural product production can be performed by metabolic engineering strategies. Selection of a host strain is based on the product. There are three ways to choose the host strain. First, the desired product can be overproduced by the host strain. Second, the desired product can be produced with low efficiency. Third, the target product is not produced by the host strain. The host strain will be generated using different metabolic engineering strategies. *E. coli* and *S. cerevisiae* are popular host strains to produce biodiesel hoewever, the efficiency is low. 'Generally recognized as safe' (GRAS) microorganisms should be used to produce food and pharmaceutical products for the safety issues. *Bacillus subtilis* and *S. cerevisiae* are well-known GRAS strains [57]. Artemisinic acid is known as an anti-malarial drug precursor and is produced in *S. cerevisiae* by introduction of heterologous pathways. Opioids were produced in *S. cerevisiae* [58]. Systems metabolic engineering can enhance the production of recombinant proteins such as artemisinic acid. Some of the target pathways are not available in microorganisms. Enzymes and metabolic pathways for the desired product should be designed by metabolic engineering. For example, lactams cannot be produced by natural pathways, and de novo pathways should be designed for lactams. Penicillin is a beta-lactam non-ribosomal peptide. Baker's yeast *Saccharomyces cerevisiae* can produce and secrete penicillin by metabolic engineering. Five genes in the benzylpenicillin pathway in *P. chrysogenum* were integrated into *S. cerevisiae*. Bioactive benzylpenicillin is then produced and secreted by *S. cerevisiae* [59].

If natural pathways of the production of the desired product is unknown, then GEM-Path, DESHARKY, RetroPath and RetroRules are used as prediction algorithm tools for metabolic pathway design. Mutations should be identified after these methods. Colorimetric assays, spectrophotometer fluorescence-activated cell sorting (FACS), or microfluidic sorting devices can be used to identify the beneficial mutation in the organism. The pathway should be optimized after the metabolic pathway is constructed in the host strain. Genome-scale metabolic simulation, plasmids, regulatory RNAs, and genome engineering are used to optimize the pathways of the host strain. Recombination-mediated genetic engineering is used to optimize pathways and produce the desired product efficiently. RecABCD system-based homologous recombination, the l Red recombination, site-specific recombination systems including Cre-lox and flippase-flippase recombinase target (Flp-FRT), zinc finger nuclease (ZFN) and CRISPR along with CRISPR/Cas are genome engineering tools [57]. The b-amyrin is a pentacyclic triterpenoid compound and was produced by *S. cerevisiae* strain engineered by CRISPRi [60].

Scale-up fermentation is an important step for biopharmaceuticals. The strain's growth performance and optimal fermentation conditions have been validated for lab-scale fermentation (0.5–30 L). After lab-scale fermentation is approved, pilot-scale fermentations (30–3000 L) and large scale production (3000–20,000 L) will be performed to see the conditions of the strain and the product of interest. Full-scale (20,000–2,000,000 L) production fermentation will be performed for the production of biopharmaceuticals. In scale-up fermentation, gradients of feed, oxygen concentrations, and maintaining the genomic stability of high-performing strains will be a challenge [27].

*DOI: http://dx.doi.org/10.5772/intechopen.96030*

## 5. Conclusions and future perspectives

Metabolic engineering is a successful method to construct efficient microbial cell factories. These methods and developments in metabolic engineering make *S. cerevisiae* an efficient and leading host strain for protein production.

## Conflict of interest

The author declares no conflict of interest.

## Author details

Seyma Hande Tekarslan-Sahin
Beykoz Institute of Life Sciences and Biotechnology, Bezmialem Vakif University, Istanbul, Turkey

*Address all correspondence to: handetekarslan@gmail.com

## References

[1] Bailey J. E.: Toward a science of metabolic engineering. Science 1991; 252:1668-1675.

[2] Deparis Q, Claes A, Foulquié-Moreno MR, Thevelein JM: Engineering tolerance to industrially relevant stress factors in yeast cell factories. FEMS Yeast Res. 2017;17(4): fox036. doi:10.1093/femsyr/fox036.

[3] Stephanopoulos N G, Aristou AA, J. Nielsen J. Metabolic engineering: principles and methodologies. 1998. Academic Press, San Diego, CA.

[4] Bailey JE, A. Hatzimanikatis ASV, Lee K, Renner WA, Tsai PS: 2002. Inverse metabolic engineering: a strategy for directed genetic engineering of useful phenotypes. Biotechnol. Bioeng.2002; 79:568-579.

[5] Nielsen J, Jewett MC: 2008. Impact of systems biology on metabolic engineering of *Saccharomyces cerevisiae*. FEMS Yeast Res.2008; 8:122-131.

[6] Bro C, Regenberg B, Forster J, Nielsen J: In silico aided metabolic engineering of *Saccharomyces cerevisiae* for improved bioethanol production. Metab. Eng.2006; 8:102-111.

[7] Sauer U: Evolutionary engineering of industrially important microbial phenotypes. Adv. Biochem. Eng. Biotechnol. 2001; 73:129-169.

[8] Sauer U, Schlattner U: Inverse metabolic engineering with phosphagen kinase systems improves the cellular energy state. Metab. Eng. 2004;6:220-228.

[9] Nevoigt E: Progress in Metabolic Engineering of *Saccharomyces cerevisiae*. Microbiology and Molecular Biology Reviews. 2008, 72 (3) 379-412; DOI: 10.1128/MMBR.00025-07

[10] Patnaik R: Engineering Complex Phenotypes in Industrial Strains. Biotechnol. Prog. 2008; 24, 38-47.

[11] Stanley D, Fraser S, Chambers PJ, Rogers P, Stanley GA: Generation and characterisation of stable ethanol-tolerant mutants of *Saccharomyces cerevisiae*. J Ind Microbiol Biotechnol 2010;37:139-49.

[12] Tekarslan-Sahin SH, Alkim C, Sezgin T: Physiological and transcriptomic analysis of a salt-resistant *Saccharomyces cerevisiae* mutant obtained by evolutionary engineering. Bosn J Basic Med Sci. 2018 ;18(1):55-65. doi: 10.17305/bjbms.2017.2250. PMID: 28954203; PMCID: PMC5826675.

[13] Teunissen A, Dumortier F, Gorwa MF, Bauer J, Tanghe A, Loïez A, Smet P, Van Dijck P, Thevelein JM: Isolation and characterization of a freeze-tolerant diploid derivative of an industrial Baker's yeast strain and its use in frozen doughs. Appl Environ Microb. 2002; 68:4780-7.

[14] Gill RT: Enabling inverse metabolic engineering through genomics. Curr. Opin. Biotechnol. 2003; 14:484-490.

[15] Raab R M, Tyo K, Stephanopoulos G: Metabolic engineering. Adv. Biochem. Eng. Biotechnol. 2005; 100:1-17.

[16] Alper H, Moxley J, Nevoigt E, Fink GR, Stephanopoulos G. Engineering yeast transcription machinery for improved ethanol tolerance and production. Science 2006; 314:1565-1568.

[17] Ko YS, Kim JW, Lee JA, Han T, Kim GB, Park JE, Lee SY. Tools and strategies of systems metabolic engineering for the development of microbial cell factories for chemical production. Chem Soc Rev. 2020

21;49(14):4615-4636. doi: 10.1039/d0cs00155d. PMID: 32567619.

[18] Yu T, Zhou YJ, Huang M, Liu Q, Pereira R, F. David F, Nielsen J: Reprogramming Yeast Metabolism from Alcoholic Fermentation to Lipogenesis Cell. 2018; 174, 1549-1558.

[19] Bassalo MC, Liu R, Gill RT: Directed evolution and synthetic biology applications to microbial systems. Curr. Opin. Biotechnol. 2016; 39, 126-133.

[20] DiCarlo JE, Conley AJ, Penttila M, Jantti J, Wang HH, Church GM: Yeast oligo-mediated genome engineering (YOGE). ACS Synth Biol 2013; 2:741-749.

[21] Si T, Luo Y, Bao Z, Zhao H: RNAi-assisted genome evolution in *Saccharomyces cerevisiae* for complex phenotype engineering. ACS Synth Biol 2015; 4:283-291

[22] Bro C, Nielsen J: Impact of 'ome' analyses on inverse metabolic engineering. Metab. Eng. 2004; 6:204-211.

[23] Jin YS, H. Alper H, Y. T. Yang YT, and G. Stephanopoulos G. Improvement of xylose uptake and ethanol production in recombinant *Saccharomyces cerevisiae* through an inverse metabolic engineering approach. Appl. Environ. Microbiol. 2005; 71:8249-8256.

[24] Lee JW, Na D, Park JM, Lee J, Choi S, Lee SY: Systems metabolic engineering of microorganisms for natural and non-natural chemicals. Nat. Chem. Biol.2012; 8, 536-546.

[25] King ZA, Lloyd CJ, Feist AM, Palsson BO: Next-generation genome-scale models for metabolic engineering. Curr Opin Biotechnol 2015, 35:23-29.

[26] Jensen MK, Keasling JD: Recent application of synthetic biology tools for yeast metabolic engineering. FEMS Yeast Res 2015, 15:1-10.

[27] Chae TU, Choi SY, Kim JW, Ko YS, Lee SY: Recent advances in systems metabolic engineering tools and strategies. Curr Opin Biotechnol. 2017; 47:67-82. doi: 10.1016/j.copbio.2017.06.007. Epub 2017 Jul 1. PMID: 28675826.

[28] Lee SY, Kim HU: Systems strategies for developing industrial microbial strains. Nat Biotechnol 2015, 33:1061-1072.

[29] Sagt CM: Systems metabolic engineering in an industrial setting. Appl. Microbiol. Biotechnol. 2013; 97, 2319-2326.

[30] Horgan RP, Kenny LC. 'Omic' technologies: genomics, transcriptomics, proteomics and metabolomics. The Obstetrician & Gynaecologist 2011; 13:189-195.

[31] Gu C, Kim GB, Kim WJ, Kim HU, Lee SY: Current status and applications of genome-scale metabolic models. *Genome Biol* 2019; 20, 121. https://doi.org/10.1186/s13059-019-1730-3

[32] El Karoui M, Hoyos-Flight M, Fletcher L: Future Trends in Synthetic Biology—A Report. Front. Bioeng. Biotechnol. 2019; 7:175. doi: 10.3389/fbioe.2019.00175

[33] Khalil A., Collins J: Synthetic biology: applications come of age. Nat Rev Genet 2010; 11, 367-379. https://doi.org/10.1038/nrg2775.

[34] Crook NC, Schmitz AC, Alper HS: Optimization of a yeast RNA interference system for controlling gene expression and enabling rapid metabolic engineering. ACS Synth Biol 2013, 3:307-313.

[35] Doudna JA,Charpentier E: The new frontier of genome engineering with CRISPR–Cas9. Science. 2014; 346,6213.

[36] Pâques F, Haber JE: Multiple pathways of recombination induced by

double-strand breaks in *Saccharomyces cerevisiae*. Microbiol. Mol. Biol. Rev. 1999;63, 349-404.

[37] Guirouilh-Barbat J, Huck S,Bertrand P, Pirzio L, Desmaze C, Sabatier L, Lopez BS: Impact of the KU80pathway on NHEJ-inducedgenome rearrangements in mammalian cells. Mol.Cell. 2004; 14,611-623.

[38] Jakočiūnas T, Jensen MK, Keasling JD: CRISPR/Cas9 advances engineering of microbial cell factories. Metab Eng. 2016; 34:44-59. doi: 10.1016/j.ymben.2015.12.003.

[39] Storici F, Durham CL, Gordenin DA, Resnick MA: Chromosomal site-specific double-strand breaks are efficiently targeted for repair by oligonucleotides in yeast. Proc.Natl.Acad.Sci.USA. 2003; 100,14 994-14999.

[40] Bao Z, Xiao H, Liang J, Zhang L, Xiong X, Sun N, Si T, Zhao H: Homology-integrated CRISPR-Cas (HI-CRISPR) system for one-step multigene disruption in *Saccharomyces cerevisiae*. ACS Synth Biol 2014, 4:585-594.

[41] Qi LS, Larson MH, Gilbert LA, Doudna JA, Weissman JS, Arkin AP, Lim WA: Repurposing CRISPR as an RNA-guided platform for sequence-specific control of gene expression. Cell. 2013; 152:1173-1183.

[42] Zhang G, Ignatova Z: Folding at the birth of the nascent chain: coordinating translation with co-translational folding. Curr. Opin. Struct. Biol. 2011;21(1), 25-31.

[43] Blockbuster Biologics 2012: sales are double that of 2006 driven by strong growth of recombinant antibodies [Internet]. 2012. Available from:https://www.pipelinereview.com/index.php/2013050950921/Press- Room/Blockbuster-Biologics-2012-sales-are-double-that-of- 2006-driven-by-strong-growth-of-recombinant-antibodies.html [Accessed: 2020.12.12]

[44] A Biopharma Year In Review — And A Look Ahead To 2020 [Internet]. 2019. Available from: https://www.biosimilardevelopment.com/doc/a-biopharma-year-in-review-and-a-look-ahead-to-0001 [Accessed: 2020.12.15]

[45] Rowland M, Noe CR, Smith DA, Tucker GT, Crommelin DJA, Peck CC, Rocci Jr ML, Besancon R, Shah VP: Impact of the pharmaceutical sciences on health care: a reflection over the past 50 years. 2012; J. Pharm. Sci.101(11), 4075-4099.

[46] Intelligence M: Global biopharmaceuticals market growth, trends & forecasts (2016-2021). 2016. Available from: https://www.mordorintelligence.com/industry-reports/ global-biopharmaceuticals-market-industry. [Accessed: 2020.12.17]

[47] Biopharmaceuticals Market - Growth, Trends, and Forecast (2020 - 2025) 2020. Available from: https://www.prnewswire.com/news-releases/biopharmaceuticals-market---growth-trends-and-forecast-2020---2025-301119584.html [Accessed: 2020.12.17]

[48] Kesik-Brodacka M: Progress in biopharmaceutical development. Biotechnology and Applied Biochemistry, 2018; 65: 306-322. https://doi.org/10.1002/bab.1617.

[49] Mitragotri S, Burke PA, Langer R: Overcoming the challenges in administering biopharmaceuticals: formulation and delivery strategies. Nat Rev Drug Discov. 2014;13(9):655-72. doi: 10.1038/nrd4363. Epub 2014 Aug 8. PMID: 25103255; PMCID: PMC4455970.

[50] Walsh G: Biopharmaceutical benchmarks 2014. Nat. Biotechnol. 2014, 32:992-1000.

[51] Gerngross TU: Advances in the production of human therapeutic proteins in yeasts and filamentous fungi. Nat. Biotechnol. 2004, 22:1409-1414.

[52] Wildt S, Gerngross TU: The humanization of N-glycosylation pathways in yeast. Nat Rev Microbiol. 2005;3(2):119-28. doi: 10.1038/nrmicro1087. PMID: 15685223.

[53] Huang M, Bao J, Nielsen J: Biopharmaceutical protein production by *Saccharomyces cerevisiae*: current state and future prospects Pharm. Bioprocess. 2014; 2(2), 167-182

[54] Fisher RM, Regenberg B: Multicellular group formation in *Saccharomyces cerevisiae*. Proc. R. Soc. B 2019; 286: 20191098. http://dx.doi.org/10.1098/rspb.2019.1098

[55] Goffeau A, Barrell BG, Bussey H, Davis RW, Dujon B, Feldmann H, Galibert F, Hoheisel JD, Jacq C, Johnston M, Louis EJ, Mewes HW, Murakami Y, Philippsen P, Tettelin H, Oliver SG: Life with 6000 genes. Science. 1996 ;274(5287):546, 563-7. doi: 10.1126/science.274.5287.546. PMID: 8849441.

[56] Choi SY, Rhie MN, Kim HT, Joo JC, Cho IJ, Son J, Jo SY, Sohn YJ, Baritugo KA, Pyo J, Lee Y, Lee SY, Park SJ: Metabolic engineering for the synthesis of polyesters: A 100-year journey from polyhydroxyalkanoates to non-natural microbial polyesters. Metab Eng. 2020; 58:47-81. doi: 10.1016/j.ymben.2019.05.009. Epub 2019 May 28. PMID: 31145993.

[57] Wang G, Huang M, Nielsen J: Exploring the potential of *Saccharomyces cerevisiae* for biopharmaceutical protein production. Curr Opin Biotechnol. 2017; 48:77-84. doi: 10.1016/j.copbio.2017.03.017.

[58] Galanie S, Thodey K, Trenchard IJ, Filsinger Interrante M, Smolke CD: Complete biosynthesis of opioids in yeast. Science. 2015 ;349(6252):1095-100. doi: 10.1126/science.aac9373. Epub 2015 Aug 13. PMID: 26272907; PMCID: PMC4924617.

[59] Awan AR, Blount B, Bell DJ, Shaw WM, Ho JCH, McKiernan RM, Ellis T: "Biosynthesis of the antibiotic nonribosomal peptide penicillin in baker's yeast." Nature communications 2017; 8:15202. doi:10.1038/ncomms15202.

[60] Meng Jie, Qiu Y, Shi S: "CRISPR/Cas9 Systems for the Development of *Saccharomyces cerevisiae.* Cell Factories." Frontiers in bioengineering and biotechnology. 2020; 8: 594347., doi:10.3389/fbioe.2020.594347.

Chapter 2

# Biotransformation of Pitanga Juice by Tannase from *Saccharomyces cerevisiae* CCMB 520

*Gustavo Monteiro, Maria Araújo, Paula Barbosa, Marcelo Mello, Tonny Leite, Sandra Assis and Amanda Sena*

## Abstract

The pitanga (*Eugenia uniflora* L.) is a native species to Brazil and widely used by Brazilian industry, mainly in food, to juice, ice cream, soft drinks, jellies and liqueurs production. The fruit contains a high concentration of anthocyanins, flavonoids and carotenoids, which make it a promising source of antioxidant compounds. The objective of this work was to produce and purify tannase from *Saccharomyces cerevisiae* CCMB 520, to apply in the integral pitanga juice and to verify its physical and chemical effects. The tannase was produced under submerged fermentation in bench bioreactor. After the fermentation process the enzyme was partially purified. The partially purified tannase was applied in the integral pitanga juice using Doehlert statistical design. The effect of the enzymatic application was analyzed by means of phenolic compounds contents and antioxidant activity. Physical–chemical analyzes were carried out to investigate the Standard Identity and Quality of the juice. The best results for partial purification were obtained by ultrafiltration. After application, the total phenolics content was 4855 mg Eq. AG/L, and for the antioxidant activity was 952 μMTrolox/L (69.41%). It has been found that it is possible by means of enzymatic treatment to improve the functional quality of the integral pitanga juice.

**Keywords:** antioxidant activity, bioconversion, *Eugenia uniflora* L., experimental design, tannin acyl hydrolase

## 1. Introduction

Tannin is a term widely used to characterize the second largest class of phenolic compounds, which, like the others, has the primordial and essential function of protecting plant tissues against attack by insects, fungi or bacteria. Tannins have a high molecular weight (500 to 3000 Da), are considered antioxidants and combine with cellulose and pectin, in addition to precipitating alkaloids and proteins [1]. These compounds occur naturally in a wide variety of vegetables, and can be found in the roots, leaves, fruits, seeds and barks. They are considered secondary metabolic products of great economic and ecological interest and have a wide value in the interactions between the plant and its ecosystem. Such compounds are responsible for the astringency of many fruits and vegetable products, due to the precipitation of salivary glycoproteins, which causes the loss of lubricating power [2, 3].

Classically, according to the chemical structure, tannins are classified into two groups: hydrolyzable and condensed. The current and most accepted classification divides the tannins into four groups (**Figure 1**): gallotannins, ellagitannins, condensed tannins and complex tannins [5]. Gallotannins are the simplest tannins and are formed by units of gallo or di-gallo esterified to a nucleus of glucose or other polyhydroxy alcohol. The molecules are usually composed of a glucose nucleus and 6 to 9 gallo groups. The most common is tannic acid [6]. Ellagitannins are esters of hexahydro-xidifenic acid (HHDP), and during its hydrolysis, the HHDP group dehydrates and spontaneously lactonizes to form ellagic acid. Condensed tannins are oligomeric and polymeric proanthocyanidins containing flavan-3-ol (catechin) or flavan-3,4-diol (leucoanthocyanins). The basic structure of complex tannins, on the other hand, consists of a unit of galotannin or ellagitannin and catechin [7, 8].

Hydrolyzable tannins can be easily hydrolyzed, either chemically or enzymatically. Tannin Acyl Hydrolase (TAH), also known as tannase (EC 3.1.1.20), is an enzyme capable of hydrolyzing tannins, leading to the release of glucose and gallic acid or ellagic acid [9]. Some are still able to perform a transesterification reaction for the production of propyl gallate [10].

TAH is a glycoprotein esterase formed predominantly by a gallic acid esterase and a depsidase. Tannase can be separated into two esterases, a specific esterase for aliphatic esters such as methyl gallate, and another depsidase that hydrolyzes depsidic bonds like m-digallic acid as shown [11]. However, the proportion between the two activities can vary according to the cultivation conditions [12]. Tannase is a biocatalyst produced by vegetables, animals, bacteria, filamentous fungi and yeast. Tannins of yeast are effective only in the decomposition of gallotannin, while bacterial and filamentous fungi are efficient in the hydrolysis of gallotannins and ellagitannins [13].

Tannase is versatile since it can be widely used in the food, pharmaceutical and chemical industries, and even in bioremediation [14]. Among the possible applications we can mention: preparation of instant teas [15], additive for animal feed [16, 17], production of gallic and ellagic acid [18, 19], synthesis of esters and effluent treatment [9, 20], beverage manufacturing (juices, beers and wines) [21] and clarification of juices [22, 23].

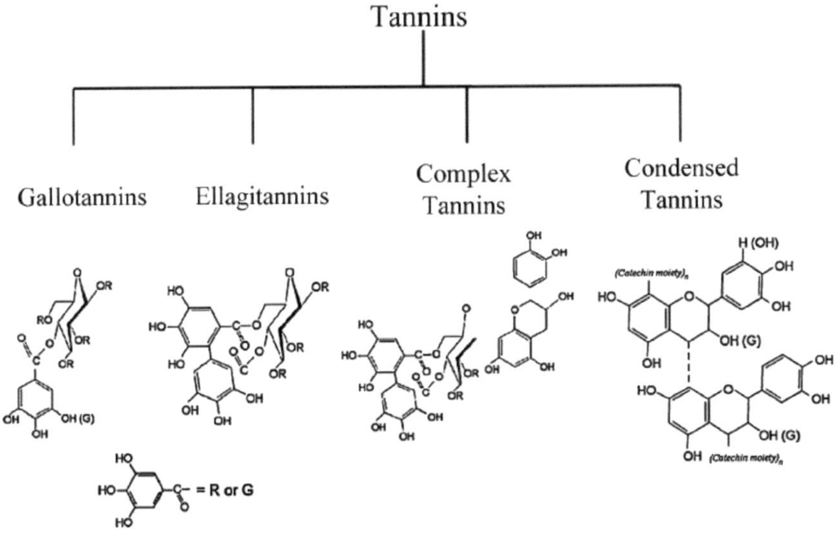


**Figure 1.**
*Main chemical structures of the tannins [4].*

The application of tannase in juices rich in hydrolyzable tannins is done to decrease the concentration of these in this food matrix, since the high content of this compound is responsible for the appearance of turbidity, bitter taste and astringency, characteristics which are often undesirable. However, the hydrolysis of gallotannins causes nutritional and sensory changes in the juice, since with the release of the gallo group occurs a retarding effect on the oxidation of ascorbic acid, also increasing its antioxidant action [24, 25].

The pitanga (*Eugenia uniflora* L.), belonging to the Mirtaceae family, is native to Brazil, specifically in the South and Southeast regions, and has adapted favorably to the edaphoclimatic conditions of the Brazilian Northeast, mainly in the State of Pernambuco, with about 300 hectares cultivated [26]. It is widely used by the Brazilian industry for the production of juice, preparation of ice cream, soft drinks, jellies and liquors because it has a high economic potential, attracting the consumer for its high concentration of metabolites such as anthocyanins, flavonols and carotenoids, which make this fruit a promising source antioxidant compounds [26, 27]. The natural antioxidants present in the diet increase the resistance to damage caused by oxidation, thus presenting a significant impact on human health [27].

Based on this information, the tannase obtained from *Saccharomyces cerevesiae* CCMB 520 was applied in this study with purpose of biotransforming the integral pitanga juice polyphenols and, in this way, modifying their biological activity.

## 2. Material and methods

### 2.1 Reagents

Tannic acid, gallic acid, bovine serum albumin and rodhanine were purchased at Sigma Aldrich (Sigma Chemical Co., St. Louis, MO, USA). All other chemicals used in the experiment were of high-quality analytical grade.

### 2.2 Microorganism and its maintenance

The yeast species *Saccharomyces cerevisiae* CCMB 520 was kindly provided by the Culture Microorganisms Collection of Bahia (*Coleção de Cultura de Micro-organismos da Bahia* - CCMB), of State University of Feira de Santana, Bahia State, Brazil. The sample was kept on plates containing Yeast Malt (YM) and left to rest in YM medium, at pH 6.8, in order to be activated; subsequently, it was incubated in B.O.D at 28 ° C for 48 hours.

### 2.3 Inoculum preparation

A 48-hours culture grown in YM medium (Merck, Darmstadt, Alemanha) was used to prepare the inoculum at pH 6.8 and 28 °C in B.O.D incubator (Cienlab, Campinas, Brazil). After the 48-hours period, culture fragments were inoculated in 0.85% saline solution to generate a suspension presenting optical density $OD_{600nm}$: 0.8 at 0.9.

### 2.4 Enzyme production and extracellular tannase obtainment

Enzyme production was performed in 7.5 L Bioreactor containing 2.5 L of submerged fermentation medium - Czapek-Dox broth (g/L) base: $NaNO_3$ (7.5), KCl (1.25), $MgSO_4.7H_2O$ (1.25), $FeSO_4.7H_2O$ (0.025), $K_2HPO_4$ $3H_2O$ (2.5), yeast extract (25) and tannic acid (150); media were sterilized at 121 °C for 15 minutes. Tannic acid (sterilized through membrane 0.45 μm) and inocolum were added to the fermentation medium after the Bioreactor cooled down to room temperature. The initial pH, fermentation time, rotation and incubation temperature, of the fermentation process, were 7, 24 h, 112 rpm and 27 °C, respectively. The fermentation broth was centrifuged (Thermoelectron, Langenser, Germany) at 1000 rpm for 15 minutes at 4 °C. The supernatant was frozen at −20 °C and used for further tests.

### 2.5 Enzyme activity and protein content

Tannase activity was estimated by using ethanolic rhodanine and tannic acid as substrate [28]. The reaction medium consisted of 250 μL substrate (0.05%, w/v) in 0.05 mol/L citrate buffer (pH 5.0) and of 250 μL enzyme extract. The substrate and the enzyme extract remained in contact for 5 minutes at 30 °C. Enzyme reaction was stopped through the addition of 300 μL ethanolic rhodanine (0.667%, w/v). After spending 5 minutes at 30 °C, the reaction medium was added with 200 μL of 0.5 mol/L potassium hydroxide in order to form a chromogen violet staining. After five more minutes at 30 °C, the obtained volume of each reaction was diluted in 4 mL of distilled water. The control tubes (enzyme extract addition at the end of the reaction) were simultaneously used. After the samples were subjected to 10 more minutes at 30 °C, the experiment proceeded in spectrophotometer (Novainstruments, Piracicaba, Brazil) at 520 nm and molar extinction coefficient was 648.15 L/mol × cm. Tannase activity (U/mL) was expressed by the amount of enzyme required to produce 1 μmol of gallic acid per minute under assay conditions. Protein content was set according to the Bradford method [29]. Bovine serum albumin was used as standard. All tests were performed in triplicate and the mean values (different from <5%) were calculated.

### 2.6 Partial purification by different methods

#### *2.6.1 Ammonium sulphate precipitation*

The crude enzyme extract was fractioned by ammonium sulphate precipitation at percentage saturation ranged of 0–20, 20–40, 40–60, 60–80% (w/v), respectively [30]. At each saturation, the solution was left to stand for 2 hours. The sample was dialyzed against distilled water for 4 hours at 4 °C and the precipitate was collected by centrifugation (5000 rpm for 20 minutes at 4 °C). The precipitates were solubilized in 0.04 mol $L^{-1}$ sodium citrate buffer (pH 5) and subjected to analysis of enzyme activity and total protein as previously described.

#### *2.6.2 Ultrafiltration membrane (30 KDa)*

The crude culture filtrate (10 mL) was added to the membrane and subsequently centrifuged at 4000 rpm for 60 minutes at 4 °C, and then the retained and permeated material were collected. The volumes obtained were separately reconstituted to the initial volume (10 mL). Soon afterwards, enzyme activity and total protein tests were performed as previously described.

#### *2.6.3 Ethanol precipitation*

The fractional precipitation followed the methodology from [31] with modifications. The solvent was cooled to a temperature of 0 °C and then added dropwise to the crude extract until you reach the desired concentrations of the same (50 to 90%, v/v). The mixture remained in contact for 1 hour at a temperature of −18 °C. After this period, the reaction medium was centrifuged at 10,000 rpm for 20 minutes at 4 °C. The precipitate was ressuspended in 0.04 M sodium citrate buffer, pH 5.0, in the same volume of crude extract added during the precipitation process. Soon afterwards, enzyme activity and total protein tests were performed as previously described. After partial purification, tannase was used in the bioconversion of integral pitanga juice.

### 2.7 Preparation of integral pitanga juice

The pitanga fruits (*Eugenia uniflora* L., 2000 g) were harvested in the orchard that is located near the Federal Institute of Education, Science and Technology of Pernambuco, Campus Barreiros, Brazil. They were collected between March and April, selected and cleaned in chlorinated water at 50 ppm for 15 minutes. Then were carried out, rinsing, pulp removal and crushing in an industrial blender. The integral pitanga juice was sifted and stored under freezing for further studies on the application of the enzyme.

### 2.8 Enzimatic biotrasformation

The statistical Doehlert [32] using two variables – partially purified tannase concentration (%, v/v) and application time (minutes) - was herein applied to investigate the best condition for antioxidant capacity increase. The enzyme extract concentration was assessed at three levels (4.5, 6.0 and 7.5%), whereas the application time was assessed at five levels (160, 180, 200, 220 and 240 minutes), which are presented in their actual values and codified in **Table 1**.

For each percentage of partially purified tannase, a control was performed, exchanging it for distilled water.

System behavior was explained through the following quadratic equation (Eq. (1)):

$$Y = \beta_0 + \beta_1 A + \beta_2 B + \beta_3 C + \beta_{11} A^2 + \beta_{22} B^2 + \beta_{33} C^2 + \beta_{12} AB + \beta_{13} AC + \beta_{23} BC + \varepsilon \quad (1)$$

Wherein: Y = experimental response, $\beta_0$ intercept, $\beta_1$, $\beta_2$, $\beta_3$ = linear coefficients, $\beta_{11}$, $\beta_{22}$, $\beta_{33}$ = quadratic coefficients, $\beta_{12}$, $\beta_{13}$, $\beta_{23}$ = interaction coefficients, A, B, C = independent variables, and ε = experimental error.

| Experiment | Partially purified tannase (%, v/v) | Application time (minutes) |
|---|---|---|
| 1 | 7.5 (0.866) | 180 (−0.5) |
| 2 | 7.5 (0.866) | 220 (0.5) |
| 3 | 6.0 (0) | 160 (−1.0) |
| 4 | 6.0 (0) | 200 (0) |
| 5 | 6.0 (0) | 200 (0) |
| 6 | 6.0 (0) | 200 (0) |
| 7 | 6.0 (0) | 240 (1.0) |
| 8 | 4.5 (– 0.866) | 180 (−0.5) |
| 9 | 4.5 (– 0.866) | 220 (0.5) |

**Table 1.**
*Doehlert matrix (real and coded) used to optimize tannase application in the bioconvertion of integral Pitanga juice.*

Each 10 mL of pitanga juice in Erlenmeyer flasks was added partially purified tannase at the proportions cited in **Table 1** and incubated in a shaker at $120 \pm 1$ rpm at 30 °C, optimal temperature of the tannase from *Saccharomyces cerevisiae* CCMB 520 [33]. After the enzymatic application was done, according to the pre-established time, the enzyme was denatured at 70 °C, for 10 minutes.

## 2.9 Physico-chemical analysis of the pitanga juice

The physical–chemical evaluation is necessary since bioconversion cannot influence the loss of quality with respect to the pre-established minimum standards for the Standard of Identity and Quality of a specific product, in this case the integral pitanga juice.

### *2.9.1 pH*

The pH was determined directly in the same with the aid of a previously calibrated pHmeter, after filtration [34].

### *2.9.2 Total soluble solids (°Brix)*

Total Soluble Solids (°Brix) was determined by a Reichert digital refractometer by dropping two drops of the sample onto the surface of the properly calibrated apparatus.

### *2.9.3 Total acidity*

A 2.5 mL sample of pitanga juice was previously homogenized and filtered in 100 mL Erlenmeyer flasks, afterwards it was diluted in 25 mL of distilled water and then stirred. Soon after, the electrode was introduced into the solution and then it was titrated with Sodium hydroxide solution (0.1 N) until the pH remained between 8.2 and 8.4 [34]. The potentiometer was previously calibrated before the analysis with pH 4 and 7 buffer solutions.

## 2.10 Total phenolics

The total phenolic content was estimated according to the Folin–Ciocalteu method [35].

## 2.11 Antioxidant activity

The antioxidant activity was assessed through the DPPH (2, 2-diphenyl-1-picrylhydrazyl) method [36].

The DPPH *radical scavenging* activity was calculated according to the equation (Eq.(2)) below:

$$\text{DPPH}\ (\%) = [(A_o - A_1/\text{A}_o)] \cdot 100. \tag{2}$$

Where $A_0$ corresponded to the absorbance of the negative control, and $A_1$ to the absorbance in the presence of the compound (sample and Trolox). Trolox was the positive control.

### 2.12 Statistical analysis

The results were analyzed in the SISVAR software - Variance Analysis System [37] and the means were compared through the Scott-Knott test at 5% probability level. In addition, the results were assessed through Analysis of Variance (ANOVA) in the Statistica Version 10.0 software (StatSoft, Inc., Tulsa, USA) [38] to find the variables presenting statistically significant effects on enzyme application ($p < 0.05$), as well as the model fitting the experimental data. All assays were performed in random order.

## 3. Results and discussion

### 3.1 Partial purification

As can be seen in **Table 2**, after the precipitation with ammonium sulphate, it was not possible to recover the activity of the enzymatic extract in the fractions of 0–20 and 60–80%. In the other fractions, it was not possible to obtain a considerable purification factor (greater than 1). Thus, it was found that the use of ammonium sulphate as a precipitating agent was not efficient in the precipitation of the target protein (tannase), since this salt may have caused the denaturation of the enzymes, under the experimental conditions evaluated.

In the precipitation using ethanol, it was found that in the 50 to 70% saturation it was not possible to verify enzymatic activity and in the concentrations of 80 and 90% a reduction in it. In purification, the most desirable is that the proteins/contaminants are decreased and the activity of the target protein is concentrated or not decreased. The use of organic solvents as a precipitating agent may have negatively influenced the activity of the enzyme, as already demonstrated by several authors [39–41]. The ethanol and ammonium sulphate might have caused denaturation through a conformational change in the enzyme tertiary structure.

| Stage | VA (U/mL) | TP (mg/mL) | SA (U/mg) | PF |
|---|---|---|---|---|
| Crude extract | 3.17 | 0.60 | 5.23 | 1.00 |
| Retained (30 KDa) | 21.080 | 0.67 | 31.66 | 6.040 a |
| Permeate (30 KDa) | 19.56 | 0.67 | 29.010 | 5.54 a |
| Ammonium sulphate (0–20%) | — | — | — | — |
| Ammonium sulphate (20–40%) | 2.41 | 0.54 | 4.46 | 0.85 |
| Ammonium sulphate (40–60%) | 1.10 | 0.80 | 1.30 | 0.24 |
| Ammonium sulphate (60–80%) | — | — | — | — |
| Ethanol (50%) | — | — | — | — |
| Ethanol (60%) | — | — | — | — |
| Ethanol (70%) | — | — | — | — |
| Ethanol (80%) | 0.19 | 0.28 | 0.66 | 0.085 |
| Ethanol (90%) | 0.27 | 0.37 | 0.72 | 0.093 |

***VA** – Volumetric activity; **TP** – Total protein; **AE** – Specific activity; **PF** – Purification factor. The experiments were performed in triplicate and the mean ± standard deviation values were presented. Values followed by the same letter did not statistically differ in the Scott-Knott test at 5% probability.*

**Table 2.**
*Partial purification of tannase from* S. cerevisiae *CCMB 520.*

In reference [42], tannase was obtained and purified from *Aspergillus niger*. and The precipitation method using ammonium sulphate (50–70%) resulted in a purification factor of 4.89. Whereas in reference [43], after partial purification of tannase obtained from *Aspergillus niger* MTCC 2425, through precipitation with ammonium sulphate (75%) were obtained a purification factor around 1.4. In reference [44] tannase from *Aspergillus nomius* GWA5 was purified after three steps, using acetone and two chromatographic processes and the authors obtained the following purification factors: 1.59 (acetone fraction), 3.21 (molecular exclusion) and 4.48 (ion exchange).

After carrying out the 30 kDa membrane separation process, was possible to verify a higher degree of compaction, resulting from the internal encrustation caused by smaller particles that were adsorbed on the tube walls, thus providing a result that characterized a partial purification (factor of purification above 1), with no statistically significant difference between the two fractions obtained (retained and permeated).

## 3.2 Biotransformation of integral Pitanga juice by partially purified tannase from *Saccharomyces cerevisiae* CCMB 520

### *3.2.1 Physico-chemical analysis*

The physical–chemical results are shown in **Table 3** and the Standard of Identity and Quality for the pitanga juice are in **Table 4**. The samples of the integral pitanga juice before and after partially purified tannase application comply with the standards required by current Brazilian legislation [45].

| Samples | pH | Total Soluble Solids (°Brix) | Total acidity (g/100 g, citric acid) |
|---|---|---|---|
| 0 | 3.40 a | 11,85 a | 1.67 a |
| 1 | 3.40 a | 12.10 a | 1.67 a |
| 2 | 3.40 a | 12.55 a | 1.57 a |
| 3 | 3.40 a | 12.25 a | 1.55 a |
| 4 | 3.30 a | 12.00 a | 1.62 a |
| 5 | 3.40 a | 10.35 a | 1.38 a |
| 6 | 3.20 a | 12.20 a | 1.66 a |
| 7 | 3.40 a | 12.20 a | 1.73 a |
| 8 | 3.40 a | 11.70 a | 1.74 a |
| 9 | 3.40 a | 12.40 a | 1.70 a |

*Sample 0: before application; Samples 1 to 9: after application. Values followed by the same letter did not statistically differ in the Scott-Knott test at 5% probability.*

**Table 3.**
*Physico-chemical parameters of integral Pitanga juice before and after application of partially purified tannase from* Saccharomyces cerevisiae *CCMB 520.*

| Legislation (BRAZIL, 2016) | Minimum | Maximum |
|---|---|---|
| pH | 2.50 | 3.40 |
| Total Soluble Solids (°Brix) | 6.00 | — |
| Total acidity (g/100 g, citric acid) | 0.92 | — |

**Table 4.**
*Standard of identity and quality for Pitanga juice.*

From the data, we can evidence that the tannase application in integral pitanga juice did not change the evaluated parameters, indicating that it would be within the pre-established national standards.

### *3.2.2 Total phenolics*

Through the results obtained for the total phenolic contents, presented here in **Table 5**, we can infer that in all tests these compounds increased when compared to their respective controls. The assay 8 (4.5% and 180 minutes) stood out statistically significantly among the others, reaching 3630 mg Eq. AG/L (285.59 mg/100 g).

The phenolic compounds are substances involved in the prevention processes of chronic diseases, including diabetes, cancer, heart disease and Alzheimer's, and knowledge about their presence in different fruit can contribute to the development of production, consumption, rural diversification and income generation [46].

In [47] after evaluating phenolic compounds in red pitanga found levels around 257 mg/100 g. Whereas in [45] found levels of 95.90 mg/100 g for the pitanga hydroalcoholic extract.

The results obtained experimentally for total phenolics were evaluated through F Test (Fisher's Test) and Analysis of Variance (ANOVA) (**Table 6**). The regression

| Assay | Total phenolics (mg Eq. AG/L) |
|---|---|
| **After application** | |
| 1 | 3630.00 ± 106,066 d |
| 2 | 4230.00 ± 318,20 b |
| 3 | 4142.50 ± 53,033 b |
| 4 | 3555.00 ± 141,42 d |
| 5 | 3842.50 ± 53,033 c |
| 6 | 3567.50 ± 159,099 d |
| 7 | 4317.50 ± 88,39 b |
| 8 | 4855.00 ± 35,36 a |
| 9 | 3955.00 ± 106.066 c |
| **Controls (white)** | |
| $C_1$ | 2655.00 ± 70.71 f |
| $C_2$ | 2467.50 ± 17.68 f |
| $C_3$ | 2630.00 ± 35.36 f |
| $C_4$ | 2830.00 ± 35.36 f |
| $C_5$ | 2467.50 ± 17.68 f |
| $C_6$ | 2642.50 ± 194.45 f |
| $C_7$ | 3205.00 ± 176.78 e |
| $C_8$ | 3567.50 ± 123.74 d |
| $C_9$ | 3242.50 ± 17.68 e |
| **Before application** | 2663.33 ± 115.47 f |

*The experiments were performed in triplicate and the mean ± standard deviation values were presented. Values followed by the same letter did not statistically differ in the Scott-Knott test at 5% probability.*

**Table 5.**
*Doehlert matrix results for total phenolics in Pitanga juice before and after application of partially purified tannase from* Saccharomyces cerevisiae *CCMB 520.*

| Variation Source | Sum of squares | Degree of freedom | Mean square | Fcal | Ftab |
|---|---|---|---|---|---|
| Regression[a] | 1324618 | 5 | 264923.60 | 9.18 | 9.01 |
| Residual | 86563 | 3 | 28854.14 | | |
| Lack of Fit | 33750 | 1 | 33750 | 1.28 | 18.51 |
| Pure Error | 52813 | 2 | 26406.25 | | |
| Total | 1411181 | | | | |

[a]*Statistically significant at 95% confidence interval. Fcal – calculated F value; Ftab – tabulated F value.* $R^2 = 0.94$.

**Table 6.**
*Analysis of variance applied to the data shown in* ***Table*** *5.*

was statistically significant (Fcal 9.18 > 9.01 Ftab) and the lack of fit indicated a good agreement (Fcal 1.28 < 18.51 Ftab) between the fitted model and the experimental data. Furthermore, the quality of the fit was also confirmed through coefficient of determination ($R^2 = 0.94$), and it implied that just 6% of the response variability was not explained by the model.

The model equation after regression, for the increase of phenolic compounds, was obtained (Eq. (3)):

$$\begin{aligned} Total\ phenolics\ (mgEQAG/L) = {} & 39442.50 \cdot (\pm 5459.69) - 4545.83 \cdot EE \cdot (\pm 805.25) \\ & + 161.11 \cdot EE^2 \cdot (\pm 49.45) - 217.19 \cdot T \cdot (\pm 40.56) \\ & + 0.36 \cdot T^2 \cdot (\pm 0.093) + 12.29 \cdot EE \cdot T \cdot (\pm 2.71) \end{aligned} \tag{3}$$

From the **Figure 2**, we found that only the interaction (positive effect) was statistically significant in the experimental field studied. **Figure 3** shows the response surface and contour curves obtained as a function of enzyme application time and tannase concentration, where it indicated that the increase in the variables under study increased the phenolic compounds. While, by decreasing the two variables, there was also an increase in phenolic compounds. This result can be seen in the positive interaction term obtained in Eq. (3) and **Figure 2**.

### 3.2.3 Antioxidant activity

Studies have shown that the consumption of fruits and vegetables reduces the risk of chronic diseases such as cancer, cardiovascular diseases and stroke [48]. This may be due to the presence of several secondary metabolites, these being related to various biological activities, including antioxidant activity.

The results of the total antioxidant activity are shown in **Table 7**, where it can be seen that test 8 (69.41%), as well as for phenolics (**Table 5**), was the one that presented values statistically superior to the other tests. We also found that all tests in the presence of the enzyme were superior to their respective controls. This demonstrates that the tannase from *S. cerevisiae* CCMB 520 acted on the compounds present in the integral pitanga juice, biotransforming them and increasing their biological activity.

The results obtained experimentally for the total antioxidant activity were evaluated by Test F and ANOVA (**Table 8**). The regression was statistically significant (Fcal 20.61 > 9.01 Ftab) and the lack of fit indicated a good agreement (Fcal 10.33 < 18.51 Ftab) between the adjusted model and the experimental data. The fit of the

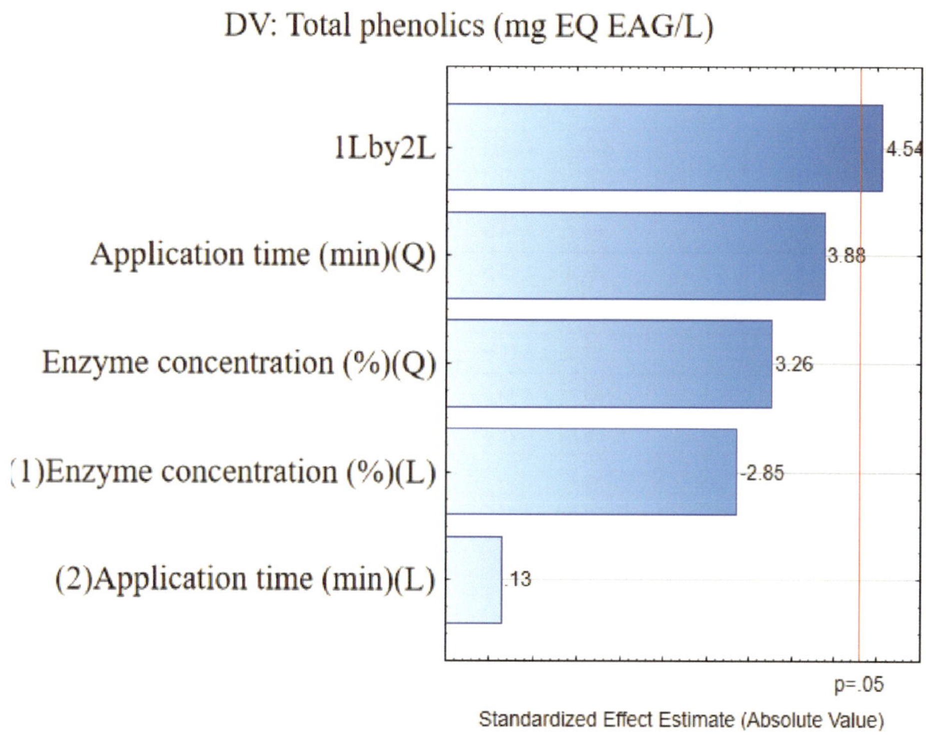


**Figure 2.**
*Pareto chart for the effects of the variables on the total phenolic content of Pitanga juice, according to statistical planning of the Doehlert design.*

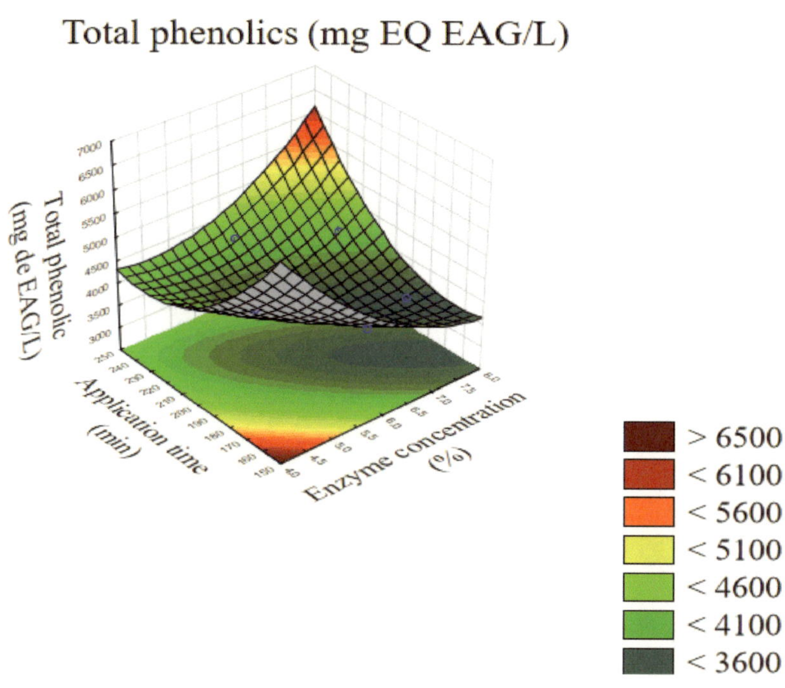


**Figure 3.**
*Response surface and contour plot to total phenolic content, according to the Doehlert design. The three-dimensional plot shows partially purified tannase concentration and application time.*

| Assay | Antioxidant activity – DPPH (%) | Antioxidant activity – μMTrolox/L |
|---|---|---|
| **After application** | | |
| 1 | 57.56 ± 1.78 d | 757.00 ± 35.56 |
| 2 | 64.96 ± 6.091 b | 803.67 ± 40.069 |
| 3 | 61.34 ± 0.59 c | 835.33 ± 4.71 |
| 4 | 64.71 ± 1.54 b | 892.00 ± 29.63 |
| 5 | 65.55 ± 0.48 b | 915.33 ± 9.43 |
| 6 | 64.71 ± 1.90 b | 863.67 ± 11.79 |
| 7 | 62.35 ± 1.19 c | 852.00 ± 23.57 |
| 8 | 69.41 ± 1.43 a | 952.00 ± 28.28 |
| 9 | 59.16 ± 0.71 d | 778.67 ± 14.14 |
| **Controls (white)** | | |
| $C_1$ | 50.00 ± 1.31 e | 607.00 ± 25.93 |
| $C_2$ | 42.017 ± 2.38 f | 448.67 ± 47.14 |
| $C_3$ | 42,10 ± 1.31 f | 450.33 ± 25.93 |
| $C_4$ | 45.46 ± 2.97 f | 517.00 ± 58.93 |
| $C_5$ | 42.27 ± 0.12 f | 453.67 ± 2.36 |
| $C_6$ | 41.76 ± 0.59 f | 443.67 ± 11.79 |
| $C_7$ | 50.42 ± 1.90 e | 615.33 ± 37.71 |
| $C_8$ | 49.07 ± 0.71 e | 588.67 ± 14.14 |
| $C_9$ | 46.97 ± 1.54 e | 547.00 ± 30.64 |
| **Before application** | | |
| | 51.26 ± 2.38 e | 632.00 ± 47.14 |

*The experiments were performed in triplicate and the mean ± standard deviation values were presented. Values followed by the same letter did not statistically differ in the Scott-Knott test at 5% probability.*

**Table 7.**
*Doehlert matrix results before and after the application of partially purified tannase from* Saccharomyces cerevisiae *CCMB 520.*

| Variation Source | Sum of squares | Degree of freedom | Mean square | Fcal | Ftab |
|---|---|---|---|---|---|
| Regression[a] | 101.01 | 5 | 20.20 | 20.61 | 9.01 |
| Residual | 2.95 | 3 | 0.98 | | |
| Lack of Fit | 2.48 | 1 | 2.48 | 10.33 | 18.51 |
| Pure Error | 0.47 | 2 | 0.24 | | |
| Total | 103.96 | | | | |

[a]*Statistically significant at 95% confidence interval. Fcal – calculated F value; Ftab – tabulated F value.* $R^2 = 0.97$.

**Table 8.**
*Analysis of variance applied to the data shown in* ***Table 7***.

model was measured by the coefficient of determination (R2), which had a value of 0.97 suggesting that 97% of the total variation in residual antioxidant activity was explained by the adjusted model. It is worth mentioning that this is the first report

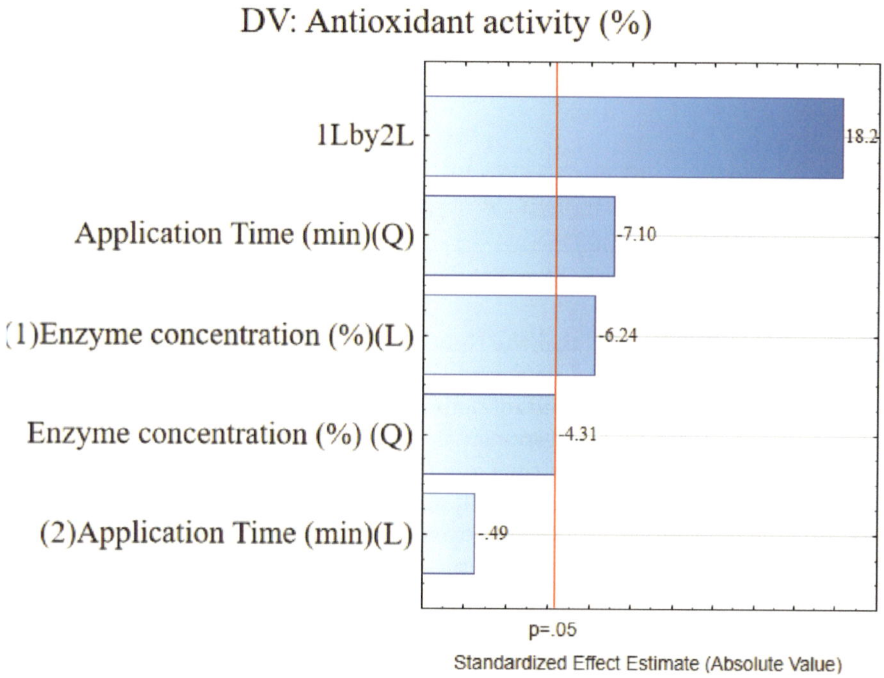


**Figure 4.**
*Pareto chart for the effects of the variables on the total antioxidant activity of Pitanga juice, according to statistical planning of the Doehlert design.*

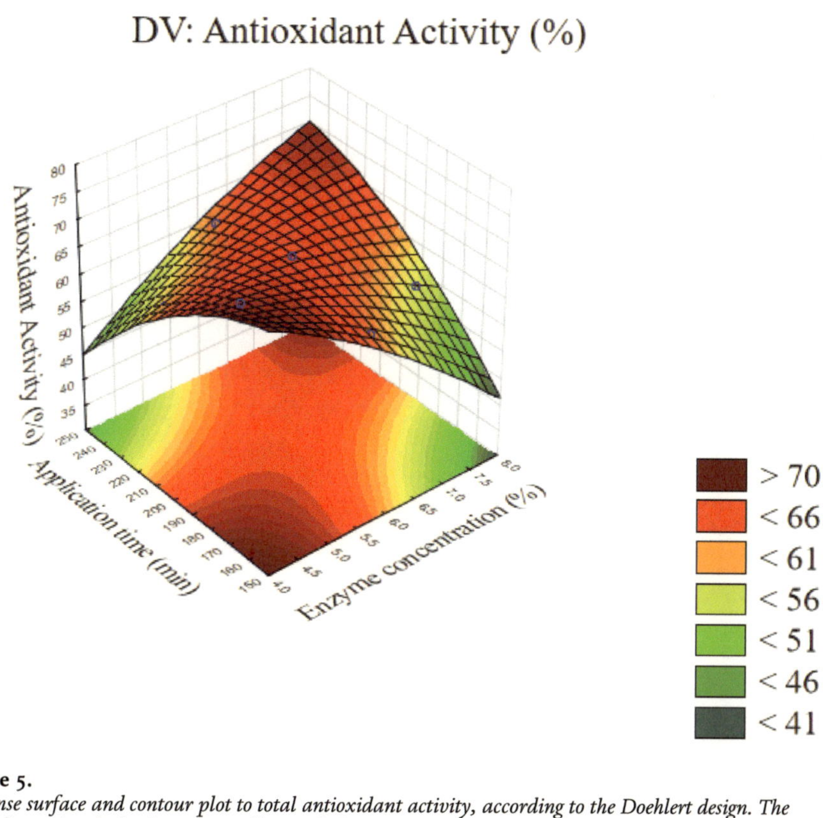


**Figure 5.**
*Response surface and contour plot to total antioxidant activity, according to the Doehlert design. The three-dimensional plot shows partially purified tannase concentration and application time.*

on the application of tannase in integral pitanga juice and its effect on total antioxidant activity and phenolic contends.

The second order equation (Eq. (4)) that describes the experimental data is presented:

$$\text{TAA}\ (\%) = 146.71 \cdot (\pm 16.29) - 22.79 \cdot \text{EE} \cdot (\pm 2.40) - 0.64 \cdot \text{EE}^2 \cdot (\pm 0.15) - 0.10 \cdot T \cdot (\pm 0.12) - 0.0020 \cdot T \cdot (\pm 0.12) - 0.0020 \cdot T^2 \cdot (\pm 0.00028) + 0.15 \cdot \text{EE} \cdot T \cdot (\pm 0.0081) \tag{4}$$

From **Figure 4**, it appears that the time in its linear term was not statistically significant in the experimental field studied.

Wherein: TAA = Total antioxidant activity.

The **Figure 5** illustrates the response surface and contour curves regarding the relationship between application time and tannase concentration. Corroborating with the data obtained for phenolic compounds, it was found that increasing or decreasing the independent variables, together, increases the response variable.

In reference [49] was evaluated samples of aqueous, ethyl acetate and butanolic extracts from pitanga fruits, where the author observed total antioxidant activity in the highest concentration (1000 μg / mL): 35.6, 86.1 and 88.7%, respectively.

Several patent filings have demonstrated the application of tannase in juices with the aim of increasing antioxidant activity. The Indiana patent application 613/KOL/2005, in [50], which describes a 37% increase in gallic acid content and an 8% increase in antioxidant activity after tannase application in the pomegranate juice. The Brazilian patent application BR 10 2015 001163–6, in reference [51] describes a total antioxidant activity of 98.20%. The results obtained by other researchers corroborate those presented in the present study. In this work, an increase in antioxidant activity of around 18.15% was possible.

Considering that the antioxidant activity is largely attributed to the presence of phenolic compounds, Pearson's correlation was calculated to verify the existence of a relationship between the two independent variables. The **Figure 6** illustrates a

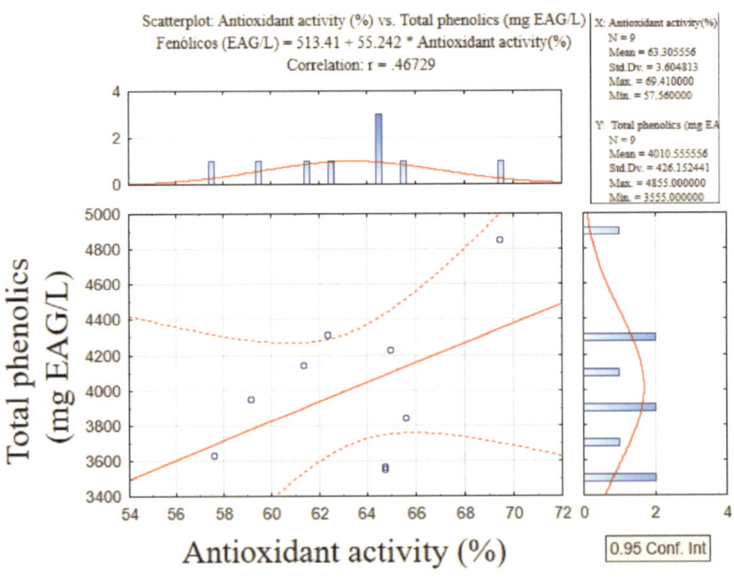


**Figure 6.**
*Pearson's correlation between antioxidant activity and total phenolics.*

moderate positive correlation between variables, by increasing phenolic compounds, antioxidant activity is increased.

The use of tannase for the release of phenolic antioxidants has become interesting for various types of food matrix. This is because most of them can release the phenolic compounds without requiring a pre-treatment such as the action of the pectinase or cellulase, or variation in temperature or pH [52]. The biotransformation of bioactive compounds is also an interesting alternative that deserves attention, since it precludes the use of toxic compounds such as organic solvents in the extraction. In these processes, bioactive compounds are obtained from natural sources by microorganisms through their secondary metabolism or by exogenous enzymatic action [53, 54]. According to [55], the bioconversion by enzyme as well as whole cell biocatalyst has tremendous importance in industry owing to escalated yields, low impurity profiles, environmental safety, and process reproducibility.

The values found after tannase application, in relation to phenolic compounds and antioxidant activity, were due to the conversion of substances present in integral pitanga juice. These data demonstrate the action of tannase obtained from *S. cerevisiae* CCMB 520 in the biotransformation of this food matrix, suggesting that the enzyme has biotechnological potential in the production of foods with better nutraceutical properties.

## 4. Conclusions

This is the first work to report application of tanase in integral pitanga juice. The purpose of the present study was to produce and apply tannase obtained from *S. cerevisiae* CCMB 520. From the results presented, we found that is possible, through enzymatic treatment, to increase the functional quality of integral pitanga juice, once there was an increase in total antioxidant activity, which is associated with an increase in total phenolic compounds.

The results suggest that the partially purified tannase of *Saccharomyces cerevisiae* CCMB 520 can potentially be used for industrial biotechnological application, as in the biotransformation of juices, to obtain a product with greater biological activity (functional property). It is worth mentioning that after the application of partially purified tannase, the juice remained with its physico-chemical characteristics within the Standard of Identity and Quality, according to the current legislation.

## Acknowledgements

We thank Federal Education, Science and Technology Institute of Pernambuco for the granted scholarships, and the National Council for Scientific and Technological Development (Grant n. 469406/2014-3) for the granted scholarships and financial support.

## Conflict of interest

The authors declare no conflict of interest.

## Author details

Gustavo Monteiro[1], Maria Araújo[2], Paula Barbosa[3], Marcelo Mello[1], Tonny Leite[1], Sandra Assis[4] and Amanda Sena[1*]

1 Federal Education, Science and Technology Institute of Pernambuco, Barreiros, Brazil

2 Federal University of Pernambuco, Recife, Brazil

3 Federal University of Paraíba, João Pessoa, Brazil

4 State University of Feira de Santana, Feira de Santana, Brazil

*Address all correspondence to: amandareges@barreiros.ifpe.edu.br

## References

[1] Brígida AIS, Rosa MF. Determinação do teor de taninos na casca de coco verde (Cocos nucifera). Proceedings of the Interamerican Society for Tropical Horticulture. 2003;47:25–27.

[2] Monteiro JM, Albuquerque P, Araújo EL. Taninos: uma abordagem da química à ecologia. Química Nova. 2005;28:892–896. DOI: https://doi.org/10.1590/S0100-40422005000500029

[3] Santos SC, Mello JCP. Taninos. In: Simões CMO, Sckenkel EP, Gosmann GJ, Mello CP, Mentz LA, Petrovick PR, editors. Farmacognosia: da planta ao medicamento. 6th ed. Porto Alegre: UFSC/UFRG; 2007. p. 885–901.

[4] Aguilar CN et al. Microbial tannases: advances and perspectives. Applied Microbiology and Biotechnology. 2007; 76:47-59. DOI: 10.1007/s00253-007-1000-2

[5] Chávez-González M et al. Biotechnological advances and challenges of tannase: An Overview, Food Bioprocess Technol. 2005;5:445–459. DOI: 10.1007/s11947-011-0608-5

[6] Pinto GAS. Produção de tanase por Aspergillus niger [Thesis]. Rio de Janeiro: Universidade Federal do Rio de Janeiro; 2003.

[7] Belmares R, Contreras-Esquível JC, Rodrigues-Herrera R, Coronel AR, Aguilar CN. Microbial production of tannase: an enzyme with potential use in food industry. Lebensmittel-Wissenschaft und-Technologie. 2004; 37:857–864. DOI: https://doi.org/10.1016/j.lwt.2004.04.002

[8] Aguilera-Carbo AF, Augur C, Prado-Barragan LA, Favelatorres E, Aguilar CN. Microbial production of ellagic acid and biodegradation of ellagitannins. Applied Microbiology and Biotechnology. 2008;23:380–400. DOI: 10.1007/s00253-007-1276-2

[9] Kumar SS, Sreekumar R, Sabu A. Tannase and its applications in food Processing. In: Parameswaran B, Varjani S, Raveendran S, editors. Green Bio-processes: Enzymes in Industrial Food Processing (Energy, Environment, and Sustainability). Singapore: Springer; 2019; p. 357–381. DOI: 10.1007/978-981-13-3263-0_19

[10] Aguilar CN, Augur C, Viniegra-González G, Favela E. A comparison of methods to determine Tannin Acyl Hydrolase Activity. BrazilianArchives of Biology and Technology. 1999;42:355–361. DOI: https://doi.org/10.1590/S1516-89131999000300014

[11] Aguilar CN, Gutiérrez-Sánchez G. Review: sources, properties, applications and potential uses of tannin acyl hydrolase. Food Science and Technology International. 2001;7:373–382. DOI: 10.1177/108201301772660411

[12] Haslam E, Stangroom JE. The esterase and depsidase activities of tannase. Biochemical Journal. 1966;99: 28–31. DOI: 10.1042/bj0990028

[13] Bhat TK, Singh B, Sharma OP. Microbial degradation of tannins: a current perspective. Biodegradation. 1998;9:343–357. DOI: 10.1023/A:1008397506963

[14] Chandrasekaran M. Tannase: source, biocatalytic characteristics, and bioprocesses for production. In: Trincone A, editor. Marine Enzymes for Biocatalysis: Sources, Biocatalytic Characteristics and Bioprocesses of Marine Enzymes. 1th ed. Reino Unido: Woodhead Publishing; 2013. 259–293. DOI: 10.1533/9781908818355.3.259.

[15] Natarajan K. Tannase: A tool for instantaneous tea. Current Biotoca. 2009; 3:96–103

[16] Neto GJS, Leal TM, Oliveira JRG, Mello MRF, Leite TCC, Sena RA.

Aplicação de tanase parcialmente purificada em teste de digestão in vitro de animais monogástricos. Brazilian Journal of Animal and Environmental Research. 2020;3:1158–1169. DOI: 10.34188/bjaerv3n3-035

[17] Sena AR, Leite TCC, Nascimento TCES, Siva AC, Souza CS, Vaz AFM, Moreira KA, Assis SA. Kinetic, thermodynamic parameters and in vitro digestion of tannase from Aspergillus tamarii URM 7115. Chemical Engineering Communications. 2018; 205:1415–1431. DOI: https://doi.org/10.1080/00986445.2018.1452201

[18] Mahmoud AE, Fathy SA, Rashad MM, Ezz MK, Mohammed AT. Purification and characterization of a novel tannase produced by Kluyveromyces marxianus using olive pomace as solid support, and its promising role in gallic acid production. International Journal of Biological Macromolecules. 2018;107(Pt B):2342–2350. DOI: 10.1016/j.ijbiomac.2017.10.117

[19] Robledo A, Aguilera-Carbó A, Rodriguez R, Martinez JL, Garza Y, Aguilar CN. Ellagic acid production by Aspergillus niger in solid state fermentation of pomegranate residues. Journal of Industrial Microbiology and Biotechnology. 2008;35:507–513. DOI: 10.1007/s10295-008-0309-x

[20] Govindarajan RK, Mathivanan K, Rameshkumar N, Shyu DJH, Kayalvizhi N. Purification, structural characterization and biotechnological potential of tannase enzyme produced by Enterobacter cloacae strain 41. Process Biochemistry. 2019; 77:37–47. DOI: https://doi.org/doi:10.1016/j.procbio.2018.10.013.

[21] Battestin V, Macedo GA. Tannase production by Paecilomyces variotii. Bioresource Technology. 2007;98:1832–1837. DOI: https://doi.org/10.1016/j.biortech.2006.06.031

[22] Prommajak T, Leksawasdi N, Rattanapanone N. Tannins in Fruit Juices and their Removal. Chiang Mai University Journal of Natural Sciences. 2020;19:76-90. DOI: 10.12982/CMUJNS.2020.0006.

[23] Silva VMA, Cruz R, Fonseca J, Souza-Motta CM, Amanda RS. Juice clarification with tannases from Aspergillus carneus URM5577 produced by solid-state Fermentation using Terminalia catappa L. leaves. African Journal of Biotechnology. 2017;16:1131–1141. DOI: 10.5897/AJB2017.15958

[24] Rout S, Banerjee R. Production of tannase under mSSF and its application in fruit juice debittering. Indian Journal of Biotechnology. 2006;5:346–350. DOI:

[25] Srivastava A, Kar R. Characterization and Application of Tannase Produced by Aspergillus Niger ITCC 6514.07 On Pomegranate Rind. Brazilian Journal of Microbiology. 2009; 40:782–789. DOI: https://doi.org/10.1590/S1517-83822009000400008

[26] Bezerra JEF, Silva Jr JF, Lederman IE. Pitanga (Eugenia uniflora L.) Série Frutas Nativas. Jaboticabal: Funep. 2000;1:30.

[27] Oliveira AL, Lopes RB. Volatile compounds from pitanga fruit (Eugenia uniflora L.). Food Chemistry. 2006;99: 1–5. DOI: https://doi.org/10.1016/j.foodchem.2005.07.012

[28] Pinto GAS, Couri S, Gonçalves EB. Replacement of methanol by ethanol on gallic acid determination by rhodanine and its impacts on the tannase assay. Electronic Journal of Environmental, Agricultural and Food Chemisitry. 2006;5:1560–1568.

[29] Bradford MM. A rapid and sensitive method for quantitation of microgram quantities of protein utilizing the principle of protein-dye-binding. Analytical Biochemistry. 1976;72:

248–254. DOI: https://doi.org/10.1016/0003-2697(76)90527-3

[30] Sivashanmugam K, Jayaraman G. Production and partial purification of extracelular tannase by Klebsiella pneumonia MTCC 7162 isolated from tannery effluent. African Journal of Biotechnology. 2011;10:1364–1374.

[31] Englard S, Seifter S. Precipitation techniques. In: Deutscher MP, editor. Guide to protein purification. 1st ed. San Diego: Academic Press; 1990. p. 285–300. DOI: 10.1016/0076-6879(90)82024-v

[32] Doehlert DH. Uniform shell designs. Applied Statistics. 1970;19:231–239. https://doi.org/10.2307/2346327

[33] Lopes LMM, Batista LHC, Gouveia MJ, Leite TCC, Mello MRF, Assis AS, Sena AR. Kinetic and thermodynamic parameters, and partial characterization of the crude extract of tannase produced by Saccharomyces cerevisiae CCMB 520. Natural Product Research. 2018;32: 1068-1075. DOI: https://doi.org/10.1080/14786419.2017.1380010

[34] Zenebon O, Pascuet NS, Tiglea P. Normas analíticas do Instituto Adolfo Lutz: Métodos físico-químicos para análise de alimentos. 1 st ed. digital. São Paulo: Instituto Adolfo Lutz; 2008. 1020 p.

[35] Singleton VL, Orthofer R, Lamuela-Raventos RM. Analysis of total phenol sand other oxidation substrates and antioxidants by means of folin-ciocalteu reagent. Methods in Enzymology. 1999; 299:152–178. DOI: https://doi.org/10.1016/S0076-6879(99)99017-1

[36] Brand-Williams, W.; Cuvelier, M. E.; Berset, C. Use of free radical method to evaluate antioxidant activity. LWT - Food Science and Technology. 1995;28: 25–30. DOI: https://doi.org/10.1016/S0023-6438(95)80008-5

[37] Ferreira DF. Sisvar: a computer statistical analysis system. Ciência e Agrotecnologia. 2011;35:1039–1042. DOI: https://doi.org/10.1590/S1413-70542011000600001

[38] Statsoft. Statistica. 7.0 by Statsoft Inc. Tulsa: Statsot, 2005.

[39] Chhokar V, Seema, Beniwal V, Salar RK, Nehra KS, Kumar A, Rana JS. Purification and characterization of extracellular tannin acyl hydrolase from Aspergillus heteromorphus MTCC 8188. Biotechnology and Bioprocess Engineering. 2010;15:793–799. DOI: 10.1007/s12257-010-0058-3

[40] Mata-Gomez M, Rodriguez LV, Ramos EL, Renovato J, Cruzhernandez MA, Rodriguez R, Contreras J, Aguilar CN. A novel tannase from the xerophilic fungus Aspergillus niger GH1. Journal of Microbiology and Biotechonology. 2009;19:987–996. DOI: 10.4014/jmb.0811.615

[41] Gonçalves HB, Riul AJ, Terenzi HF, Jorge JA, Guimarães LHS. Extracellular tannase from Emericella nidulans showing hypertolerance to temperature and organic solvents. Journal of Molecular Catalysis B: Enzymatic. 2011; 71:29–35. DOI: 10.1016/j.molcatb.2011.03.005

[42] Al-Mraai STY, Al-Fekaiki DF, Al-Manhel AJA. Purification and characterization of tannase from the local isolate of Aspergillus niger. Journal of Applied Biology & Biotechnology. 2019;7:29–34. DOI: 10.7324/JABB.2019.70106.

[43] Nandi S, Chaterjee A. Extraction, partial purification and application of tannase from Aspergillus niger MTCC 2425. International Journal of Food Science and Nutrition. 2016;1:20–23.

[44] Farag AM, Hassan SW, El-Says AM, Ghanem KM. Purification, characterization and application of

tannase enzyme isolated from marine Aspergillus nomius GWA5. Journal of Pure and Applied Microbiology. 2018; 12:1939–1949. DOI: http://dx.doi.org/10.22207/JPAM.12.4.30.

[45] Brasil. Ministério da Agricultura, Pecuária e Abastecimento. Secretaria de Defesa Agropecuária. Instrução Normativa N° 37, de 1° de outubro de 2018. Diário Oficial da União: Edição 194, seção 1, 23, Brasília, DF, 08 de out. de 2018.

[46] Souza AG, Facina AC, Saraiva F. Compostos bioativos e atividade antioxidante em frutas nativas do Brasil. Saraiva Agrotrópica. 2018;1:30, 2018. DOI: 10.21757/0103-3816.2018v30n1p73-78.pdf

[47] Lima VLA, Mélo EA, Lima DES. Fenólicos e carotenóides totais em pitanga. Scientia agrícola. 2002;59:447–450. DOI: https://doi.org/10.1590/S0103-90162002000300006

[48] Yeum KJ, Russell RM, Krinsky NI, Aldini G . Biomarkers of antioxidant capacity in the hydrophilic and lipophilic compartments of human plasma. Archives of Biochemistry and Biophysics. 2004;430:97–103. DOI: doi: 10.1016/j.abb.2004.03.006

[49] Pessanha FF. Eugenia uniflora L. (Myrtaceae): caracterização e avaliação dos compostos fenólicos, da vitamina C e da atividade antioxidante dos frutos da pitangueira [Dissertation]. Campos dos Goytacazes: Universidade Estadual do Norte Fluminense Darcy Ribeiro; 2010.

[50] Indian Institute of Technology. Enzymatic debittering of pomegranate juice. India. Indian Patent Application Number 613/KOL/2005.

[51] Amanda Reges de Sena. Suco de Jamelão e processo para fabricação de suco de jamelão utilizando extrato enzimático a base de tanase. Brasil. Depósito de pedido de patente n° BR 10 2015 001163–6. Revista de Propriedade Intelectual, n. 2405, 07 de fevereiro de 2017.

[52] Junior JVM, Teixeira CB, Macedo GA. Biotransformation and bioconversion of phenolic compounds obtainment: an overview. Critical Reviews in Biotechnology. 2015;35:75–81. DOI: 10.3109/07388551.2013.803020

[53] Martins S, Mussatto SI, Martínez-Avila G, Montãnez-Saenz J, Aguilar CN, Teixeira JA. Bioactive phenolic compounds: production and extraction by solid-state fermentation. A review. Biotechnology Advances. 2011;29:365–73. DOI: https://doi.org/10.1016/j.biotechadv.2011.01.008

[54] Puri M, Sharma D, Barrow CJ. Enzyme-assisted extraction of bioactives from plants. Trends in Biotechnology. 2012;30:37–44. DOI: https://doi.org/10.1016/j.tibtech.2011.06.014

[55] Dhiman S, Mukherjee G, Singh AK. Recent trends and advancements in microbial tannase-catalyzed biotransformation of tannins: a review. International Microbiology. 2018; 21: 175–195. DOI: https://doi.org/10.1007/s10123-018-0027-9

Chapter 3

# Understanding the Mechanism of Action of Indigenous Target Probiotic Yeast: Linking the Manipulation of Gut Microbiota and Performance in Animals

*Shakira Ghazanfar*

## Abstract

The gut associated microbiota of animal plays crucial rule in the conversion to accessible nutrients for improve animal health and well-beings. Probiotic yeast (PY) is commonly use to manipulate the gut microbial balance by inhibits the disease-causing microbes and increase the number and function of desirable microbes. PY produce many fermentation metabolites, intercellular effectors, minerals and enzymes that make it an idea nutritive feed supplement for ruminants. The mode of action of the PY is depends on the animal biological inheritance, breed, managemental condition and microbial feeding type. Therefore, PY must formulate using same ecological origin, alone with desirable target; as it would be more compatible with gut ecoysytem and would yield maximum outputs as compare to non-target or foreign probiotic (FP). Therefore, for development of the Indigenous Target Probiotic (ITP), the isolation source must be same ecological region with desirable target like improve animal health and productivity. In the situation of the increase food storage around the world, ITP may provide a useful feed supplements to improve the food production in cost effective manner as compare to FP. Probiotic effectiveness is considered to be population/breed/target specific due to difference in the feed intake, change gut microflora, different food habits and different host-microbial interactions. In this chapter, we will highlight the preparation of the ITP yeast and its mode of action on animal gut microbiota.

**Keywords:** indigenous target probiotic (ITP), *Saccharomyces cerevisiae*, mode of action, gastrointestinal tract, fiber digestion

## 1. Introduction

Probiotic are the live microbial feed supplements which provide the beneficial impact on the host by producing the useful metabolites [1]. Many probiotics have been available in the market for improving animal and human health in safe and healthy way. The commercially available probiotic product contains mostly lactic acid bacteria (*Lactobacillus plantrum, L.casei etc.)* and yeast (*Saccharomyces cerevisiae*) strains [2]. The beneficial impact of present probiotics is often limited and do not provide equal affects to each host. The positive impact of the probiotic product is based on the site of action, its dose, the stability/viability of the microbial strain; host genome and its environmental condition and health [3]. Mode of action of the microbial strains is one of the majors determines of the probiotic yeast usefulness. Latest molecular methods must be used for identification of the unique microbial strains for development of target-based probiotic yeast. During the last decades, probiotic yeast (*Saccharomyces cerevisiae)* has been extensively used as ruminant health promoter [4]. The beneficial outcomes from probiotic product mostly depends on the host and microbial interaction, therefore, pre-plan steps must follow for isolation of the best performance (target) microbial strains for development of the unique/true animal probiotic yeast [5]. Ruminants have a unique microbial flora which is responsible for breakdown of the fibrous and non-fibrous feed particles. The number and function of the gut microbes is highly affected by biochemical and microbial properties of the rumen [6]. The gastrointestinal tract microbial flora has a crucial role on upgrade nutrient utilization and feed digestion leads to the improve animal production and health status. Animal eat different types of feed (high energy & low energy), that determine the number and function of the microbes in the gastrointestinal tract. The gut microbiota is highly changeable due to the addition of useful microbial feed supplements in safe and healthy way as compare to any antibiotic [7]. Animal blood profile also plays an important role in the animal health and its production performance. PY brings changes in the concentration of rumen volatile fatty acid (VFAs) propionate, butyrate and valerate leads to the reduced synthesis of triglyceride and cholesterol in the liver cells and might be change the lipid profile in blood. These polysaccharides reduce the total cholesterol of serum in ruminants. Therefore, the blood chemistry and the fecal microbiota must be manipulated for better animal health and performance. Literature showed that the

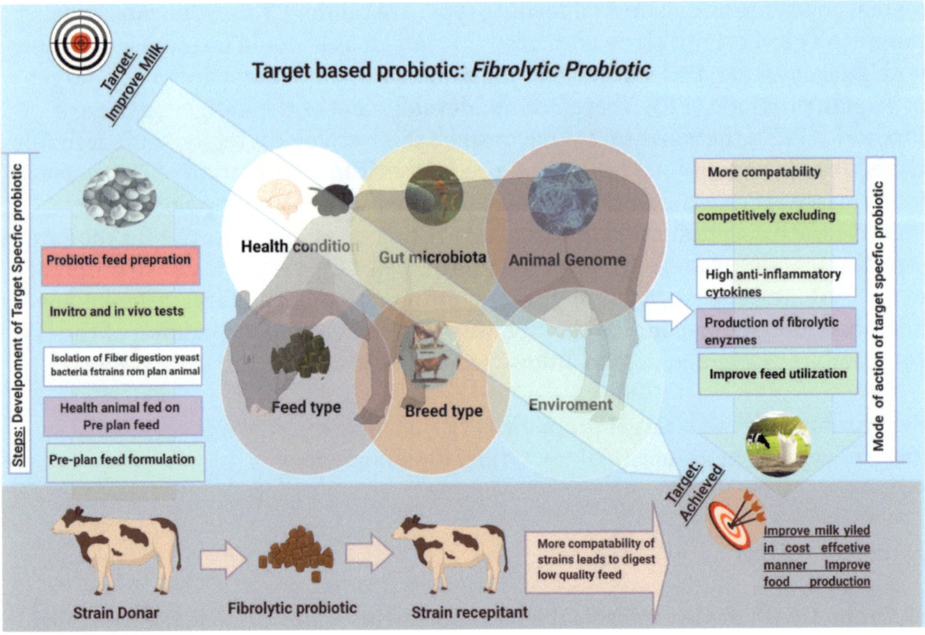


**Figure 1.**
*Representative scheme of development of target-based probiotic (TBP): The right side covers the main steps involving in the preparation of the TBP, the internal part covers the legalistic evidence of the interrelationship between, host and microbes. The left side covers the mechanistic activity of the TBP; including the improve gut microbial balance which leads to the improve feed digestion resultantly improve host health and production in cost effective manner.*

microbial diversity of the animal GIT is very important in feed digestion processes. Ruminants has a big anaerobic chamber/vat called rumen. Inside rumen, three main microbial species, i.e. bacteria, fungi and protozoa are present for feed digestion. Rumen microbial flora digests the lignocellulosic biomass and release the energy (VFAs) for animal use. Rumen microbial flora are animal best friends. If required specialized gut microbial flora are not present, the food digestion process can be shut down and death of the animal can occur. For colonization of the best microbial flora inside the rumen, we must formulate animal feed after clear understanding of the rumen ecosystem, and host genetic (**Figure 1**) [8, 9].

In the situation of high animal feed cost, we must identify the cost-effective probiotic by using the concept of ITP to improve poor quality feed into high quality milk and meat. We had already given the concept of indigenous probiotic yeast our previous book chapter [31]. A clear understanding regarding the proposes guidelines to develop the ITP to improve gut microbiota resultantly improve milk and meat production. This book chapter will discuss the identification of the microbial strain from local ecological breed and its mode of action for preparation of target based probiotic products. We will also support our concept of ITP with our lab conducted experiments.

## 2. Yeast: promising microbe for development of target probiotic for animal use

Yeast is a very useful microorganism with broad range of industrial application, because of their unique genetics and physiology. Yeast cells have many useful metabolites (protein, carbohydrate, vitamins; vitamin B6, thiamin, biotin, riboflavin, nicotinic acid and pantothenic acid and minerals; zinc and magnesium) [10]. The utilization of the naturally prepared yeast would be accelerated in coming years due to the nature-oriented mind set of the consumers. Therefore, research on the isolation of the nutritious rich yeast strains for preparation of probiotic product has rapidly increased [11, 12]. Yeast is an important single cell microorganism, belongs to fungus family and it multiplies by cell division. The genetics and physiology of the yeast are very unique, and, therefore, a broad range of research work in biological sciences is being carried out on this microbe. The yeast cell size is composed of 5 × 10 μm and the size of the baker's yeast genome is 12.1 Mb containing 16 chromosomes and 5400 coding genes approximately [13]. Members of the order Saccharomycetales are mainly used for the animal probiotics when serves as reliable and economical source of essential amino acids, vitamins, carbohydrates, and minerals from yeast cell. Thiamin, Riboflavin, Niacin and Biotin are present in yeast [14]. The antagonistic ability of the yeast to block bacterial pathogenicity is also makes its very useful [15]. Yeast cell has competition for nutrients, pH changes in the medium, high concentrations of ethanol production, secretion of antibacterial compounds and release of antimicrobial compounds are major antagonistic steps. Yeast cell has many useful fermentation metabolites (protein, vitamins, carbohydrates) which makes it important microbial feed supplement. Yeasts are naturally present (1.3 X $10^5$ yeasts ml-1) inside the rumen fluid [16]. Literature showed that, yeasts (*Sac. Cerevisiae*) are not significant members of the rumen microbial flora, but mostly, entering inside the rumen with fibrous feed [17]. Therefore, we claim that the viable yeast rich diet can improve the its numbers and function inside the rumen. Now a days, *Saccharomyces cerevisiae* (live yeast) has been extensively used as animal probiotic to improve milk production and its composition. Many researchers have given different types of conclusion related to the mode of action of yeast and its impact on host animal. Mostly all researchers agreed that the improved live bacterial count inside rumen is the most reproducible impact of PY [18–24]. Based upon a research, it is being hypothesized that probiotic effectiveness is considered to be population-specific due to differences in the feed, gut microflora composition, food habits and host-microbial interactions. We can isolate and identify the target yeast strain from animal gut and can used that strain for preparation of the animal probiotic yeast.

## 3. Probiotic yeast for neonatal and growing ruminant diet

The role of the probiotic yeast in dairy animal is well studied [25]. They have been extensively used for improve milk yeast and its composition in cost effective manners. The benefits to cost ratio of probiotic yeast is 4:1 in dairy animals. They have also used as preventer against digestive problems, and rumen acidosis.

The main target of the PY used in new born ruminate diet are; (a) improvement in the rumen maturation; (b) stop the pathogenic bacterial growth; (c) establishment of the normal growing animals like microbial flora [26–28]. Microbial based feed can improve the rumen development during the growing phase of the dairy animals. The new born gut is sterile and have no germ [29]. After 6 months of age the rumen is colonized with diverse microbial flora. PY provides beneficial metabolites and enzymes like thiamine for fast growth of the fungi. The poor fungal growth of the animal fed on PY might be due to the low production of thiamine [30]. At the same time, the animal plays an important role in the maximum colonization of the beneficial microbial population [31]. If there is any imbalance bacterial species, it would result in digestive problems and leads to the economic loss. The establishment of the useful bacterial strains results in the development of strong and balanced rumen which resultantly strong immunity and health condition [32, 33]. PY provide the improve the rumen maturation and its microbial flora is also in strong balance. PY provide the useful bacterial species for feed digestion, like cellulolytic bacterial species and ciliate protozoa [34]. The balance in rumen microbial flora plays a crucial role in feed utilization and could result in better animal productivity [35]. PY remove oxygen from rumen and provides a more anaerobic environment for its growth of key beneficial microbial groups [36]. The newborn gut can easily be modulating by PY. The new born key beneficial microbial *Bacteroides-Prevotella* and the *C. coccoides -E. rectale* group easily be grown with PY presence by removing the oxygen inside rumen [17]. Under filed conditions, crossbred animal are usually underfed, which results in deficiencies of certain nutrients and ultimately reflected in the levels of certain biochemical constituents Literature showed that the use of PY may enhance the blood and fecal biomarkers leading to to improved health status in dairy animals [37–40].

## 4. Manipulation of ruminal gut microbiota by target probiotic (*Fibrolytic probiotic*)

For clear understanding of the ruminal gut microbiota using latest genomic methods to get useful information for preparation of specific probiotics. The ruminants feed consists of concentrate, silage, seasonal fodders etc. There diet mostly contains cellulose, hemicellulose starch and water-soluble carbohydrate. The rumen microbes play an important role in feed digestion. The animal feed is digested inside rumen and then energy is released for animal use. Cow and its microbes are mutually benefiting each other (**Figure 2**). Rumen is the first and the largest anaerobic chamber of the cow GIT. The temperature inside the rumen chamber is between

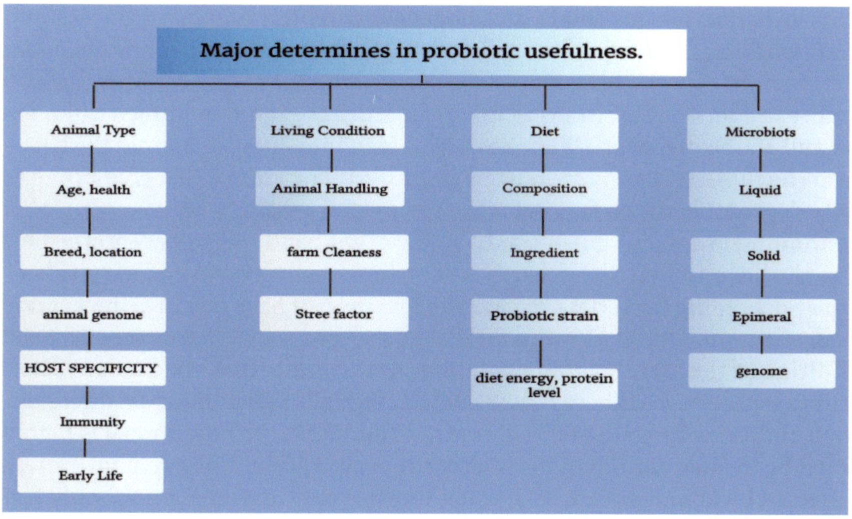


**Figure 2.**
*Major factors effects on the mode of action of probiotic.*

38 to 41 °C, with 6-7 pH (depends on feed type). There are three different types of microbes present inside the rumen including, bacteria, fungi and ciliated protozoa [41–44]. The location and size of the rumen microbes depends on the feed formulation and host genetic. Mostly, bacteria are associated with fibrous feed particles; fungi, protozoa [45, 46]. Some are freely living and some are bound with rumen mucous membrane. 1 ml of the rumen is composed of 109 to 1010 per ml bacteria with 200 different species, $10^4$ to $10^6$ per ml protozoa with 20 different species, and $10^3$ per ml fungi with 20 different species [47]. The rumen bacteria are gram negative 1-2 micrometer in size and cocci, and rod shaped mostly. Rumen bacterial are mostly non-spore producing, facultative anaerobes. 1- 5 % of the bacterial cells in rumen are cellulose digesters [48]. The rumen fungi (gut fungi) also play an important role in fiber digestion by stimulating growth of fibrolytic bacteria [49]. The rumen microbial features are heritable; moreover, animals age, feed and genome plays an important role in the microbial colonization. The composition of the diet describes the type of gut microbial species [50]. Therefore, the rumen microbiota can be manipulated by using the yeast-based probiotic to obtained the useful products. The feed must be targeted for modulating the rumen microbiota (**Figure 3**).

The modulation of the rumen microbiota is mostly for the enhanced colonization of the fiber digesting microbiota [35, 36]. Literature showed that, animal diet has an important role in the manipulation of the rumen microbiota. Low amount of fibrous feed builds up fast working microbes (fibre-degrading

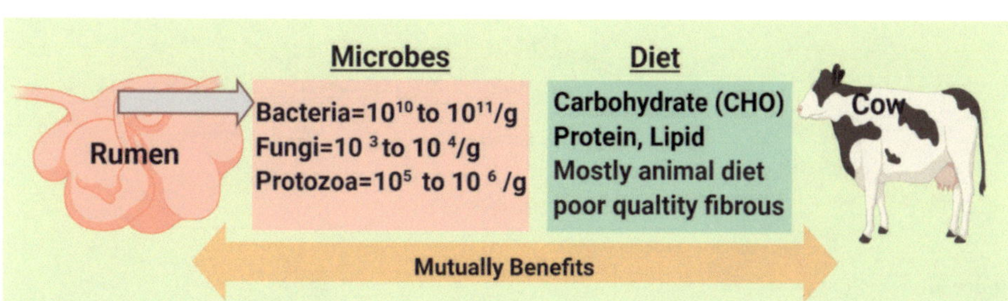


**Figure 3.**
*A scheme describing the mutually benefits between host microbes.*

Butyrivibrio fibrisolvens and F. succinogenes) and high amount of fibrous build up slow working fiber degrading microbes *(M. elsdenii, S. bovis, S. ruminantium, and P. bryantii)*. On the fibre mat of the rumen, the slow working fibre digestion microbes reside. The fast working microbes are present on the rumen fluid, for sugar and starch digestion. Microbes digest feed into end product so, the balance in the rumen microbiota must be improved. The animal diet containing the rapidly degradable starch facilitates the removal of ciliated protozoa populations (Entodinium) from the rumen fluid. On the other hand, high concentrate diets lead to the low ruminal pH which more detrimental to growth and survival of the fiber degrading bacterial species. The low pH can have negative impact of the growth of ruminal fungi [36–47]. Similarly, zoospores by Caecomyces decreases by addition of the more soluble sugar [35]. At the same time, the best growth of the fungal spores occurs between 39-40 °C. High-fibre diet might facilitate the growth of diverse fungal species in rumen. Therefore, the host animal is highly affected by the diet formulation its nutrient composition. Rumen fungi growth is alo affected by animal breed, its age and breed type. Gut fungi are the only fungi for which no oxygen is required for completion of their life cycle and the presence

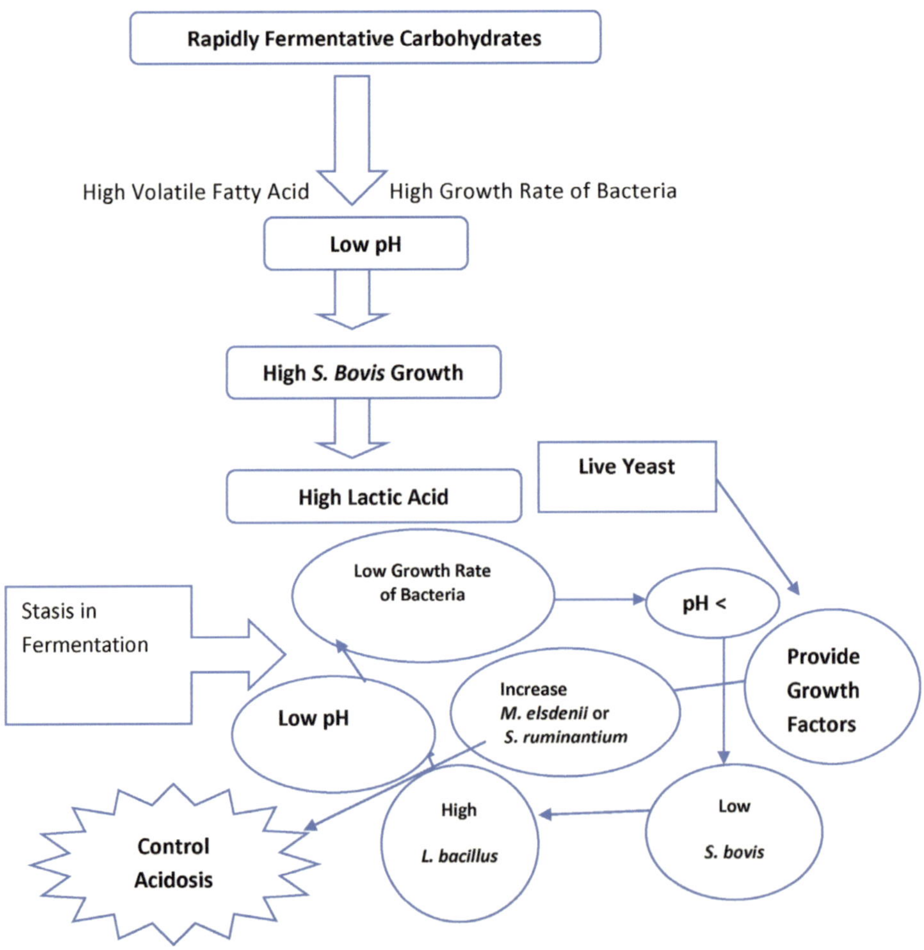


**Figure 4.**
*Potential mechanisms of microbial ruminal acidosis: This figure suggested that, the live yeast supply different growth factors (amino acid, peptides, vitamins and organic acids). These growth factors have the knock-on impact of increases the stimulation and metabolism of lactic acid utilizing anaerobic bacteria, such as M. Elsdenii or S. ruminantium (that control the acidosis). Yeast cells has a affinity for sugar which outcompete S.bovis for the utilization of sugar.*

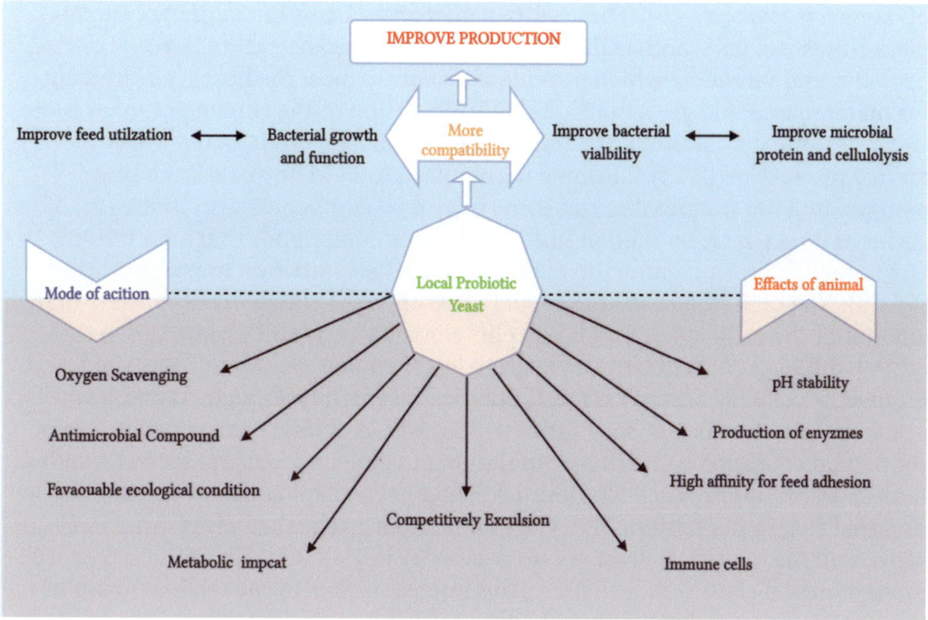


**Figure 5.**
*A proposed flowsheet to explain mechanistic pathway of IPY: Steps involved in the mode of action of PY and its impact on animal.*

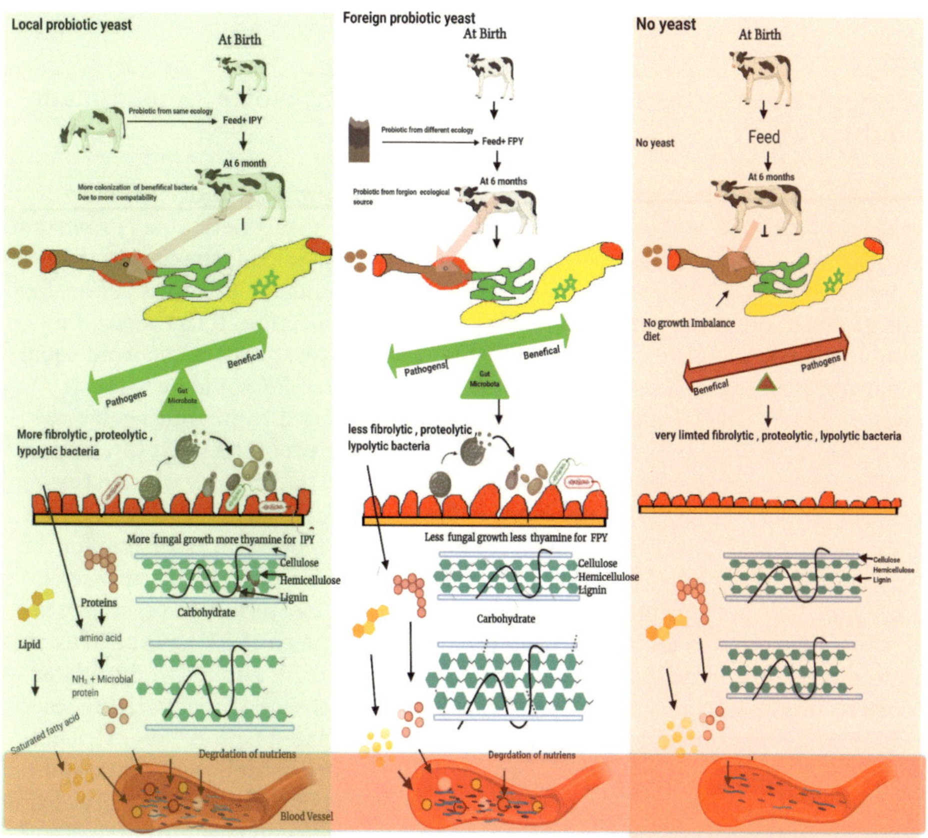


**Figure 6.**
*A simple scheme proposed to explain mode of action of probiotic yeast in gut: IPY improve carbohydrates, protein and lipid digestion rate by improving the production of cellulolytic, hemi- cellulolytic and proteolytic and lipolytic bacteria and fungi as compare to FPY and no yeast animal.*

of oxygen is toxic [35–56]. These rumen microorganisms can degrade complex plant fibers and polysaccharides and produce volatile fatty acids (VFAs), microbial proteins, and vitamins, which provide nutrients to meet the host's requirement for maintenance and growth [35, 36]. Manipulation of the rumen gut microbiota could be done for obtaining the required fermentation product and improve animal production [57, 58]. Rumen manipulation could be made by change/ manipulated the feed intake, and some microbial supplement/probiotic [35, 36]. As far as lipid part is noted that lipid are organic compounds that are insoluble in water but soluble in organic solvents. Fat and oil are nutrition important lipids [57–59]. The high forages diet leads to high rumen pH which in turn results in high amount of the cellulolytic and hemicellulolytic bacteria and protozoa, On the other hand high concentrate diet leads to lower rumen pH which results in lower number of cellulolytic and hemicellulolytic and amylolytic bacteria and lower number the rumen protozoa (**Figure 4**) [59, 60]. Probiotic change rumen environmental condition through manipulation of rumen microbiota for our required fermentation end product. The animal feed must be kept constant to build up the required rumen microbiota [61]. Cow microbiota established after some weeks of birth, and the microbial diversity increases day by day [62]. The animal feed, the managemental condition, genetics plays important role in the establishment of the animal gut microbiota [63, 64]. Once established, if the feed and the life style same, the number and function of the rumen microbiota mainly same throughout life. But we can manipulate the GIT microbiota for our own purpose. If we isolate the fiberlytic yeast strains from the rumen, we can prepare the best and unique probiotic yeast for improve animal feed digestion (**Figures 5** and **6**).

## 5. Prepration of indigenous probiotic yeast: right choice for maximum outcomes

The gut microbiota can digest the animal feed and produce nutrients for improve host health and well beings. Animal feed and host genetics play important role in shaping and composition of gut microbiota [18]. Same is the case of the rumen microbiota, which is highly variable and is depended on various factors like animal breed, physiology, feed type and geographical location. It has been commonly accepted that commercially available probiotic yeast may not showed equal impact to all animal breeds [65, 66]. The compatibility of PY could be variable among animals. The local prepared yeast probiotic isolated from same ecological niche may have more beneficial impact than any exotic probiotic yeast [3]. The local isolated probiotic yeast may have fast adaptability and colonization in the local rumen ecosystem [24]. The origin of the probiotic strain determines the best prepared probiotic product. The strain selection is the most important step for the development of right probiotic for animal. Being precise during the strain's selection could yield positive outcomes from the probiotic. The probiotic yeast may use for the rumen microbial manipulation [67]. Different types of PY have been used for improve animal health and production [7–68]. Some PY strains produced beneficial results in animals while others did not. The difference of that variable results of PY may be explained by different host and PY associated factors [69–71]. These factors are; animal age, breed, sex, feeding dose, PY strains isolation source and some unknown factors [3]. The major factors might be the low compatibility of the exotic probiotic yeast strain with animal having diverse biological inheritance and gut microbial composition. The right probiotic strain should be novel, so we must use latest molecular methods to isolate the target specific/local isolated microbial strains. The local isolated and molecular identified probiotic strains may have more impact on local animals in cost effective manners. The probiotic are species specific by targeting the indigenous strains and local dairy farms can get the cost-effective probiotic product for improve milk production and composition.

The main steps involved in the preparation of the breed specific probiotic yeast are as following [3].

- Pre-plan ruminate diet for isolation of probiotic yeast
- Identification of yeast strain based on the molecular techniques
- Probiotic potential of selected yeast strains
- In vitro probiotic potential
- Safety assessment/In-vivo animal model

## 6. Mode of action of the IPY Vs FPY inside the rumen and post-ruminal GIT

The first mode of action of the probiotic yeast is competitive exclusion (CE) [27]. The CE is a probiotic mode of action that involves the colonization of the beneficial microbial strains to GIT tract to reduce the addition of disease-causing microbial flora [18–74]. The ability of probiotic yeast cell to fight with other useless microbial flora can improve growth and function of beneficial microbial flora. The IPY has the indigenous strain, which has the advantage that it drives from animal of interest (Cow). IPY has an environmental modification capability. The concept of co-evolution of host microbial has been seen in case of IPY mode of action. The local strain gains an advantage because of its ability to adjust/modify itself in new environment by producing the antimicrobials e.g (lactic acid) to make its less suitable for its competitors. The FPY has the foreign origin strain, which has the less environmental modification capability less, competition for available nutrients, and mucosal adhesion sites. Second mode of action of the PY is reported as a good pH stabilization. Rumen microbial flora can work under stable pH [75]. Rumen pH is highly affected by animal feed intake and its composition. Ruminants eat different types of feed, like high energy concentrate diet, fodder, and silage. These types of feed have a quick impact on rumen pH. If rumen pH is not stable, the animals may have different types of metabolic diseases [76]. Literature showed that PY has a stabilizing effect on the rumen pH [77, 78]. Some studies reported a rise rumen pH when animal was fed on diet with high energy supplemented with PY. Sometimes, the increased pH might be due to the decreased VFAs inside the rumen [3–79]. The lower pH leads to the rumen acidosis, PY can prevent the acidosis condition of the dairy animals [7]. The third proposed mechanism is that yeast cell provides the anaerobic condition inside rumen by removing the oxygen thus facilitated the useful feed digestion microbes [35, 36]. The main microbial flora are bacteria fungi and protozoa. These microbial species have a fiber digestion role by secreting the cellulase and hemicellulase enzymes. Fiber is the main part of the ruminant diet. Therefore, fiber digestion, nature blessed them with unique fibrolytic digestion bacteria (*Fibrobacter succinogenes, Ruminococcus flavefaciens, Ruminococcus albus*), fungi (*Necallimastix*) and protozoa. That complex fibrolytic microbes catalyze the rumen fiber digestion and improve feed intake. The yeast supplementation provides the useful metabolites which stimulate the growth of fiber degrading bacteria [18–47].

## 7. Experimental proofs: who is better; indigenous or foreign microbe as animal probiotic?

### 7.1 Experiment: impact of probiotic yeast on blood fecal biomarkers in dairy heifers and growing animals

Based upon the above discussion, we have conducted two research experiments on dairy animals by using the IPY concept to improve the gut health. In experiment 1, eight dairy heifers (87 ± 5 kg and 6–7 months) were divided into two equal groups (control n = 4 and probiotic n = 4) [80]. Control group animals fed on NRC recommended diet and probiotic group animals fed control diet FPY (Yea-Sac$^{1026}$; 5 g/animal). After 120 days results showed that the FPY significantly affected the serum glucose, and urea levels in dairy heifers [24].

| Items | Feeding regime | | p-Value |
|---|---|---|---|
| | Control[2] | FPY[3] | |
| **Urea (mg/100 ml)[1]** | | | |
| Before treatment[4] | 30.10 ± *0.711 | 31.14 ± 0.974 | 0.012 |
| After treatment[5] | 33.34 ± 0.432 | 29.23 ± 0.494 | 0.01 |
| **Glucose (mg/100 ml)** | | | |
| Before treatment | 62.67 ± 4.04 | 60.86 ± 2.80 | 0.605 |
| After treatment | 63.31 ± 2.60 | 65.47 ± 2.84 | 0.600 |

*[1]n = 4 per treatment.*
*[2]Control feed without yeast.*
*[3]Probiotic feed compose of control feed supplemented with 2.5×10 $^{07}$ cfu/g commercially available probiotic yeast (Yac-Sac$^{1026}$) at the rate of 5 g per animal/day * ± Standard error of the mean.*
*[4]Before treatment (day 0).*
*[5]After treatment (day 120).*

**Table 1.**
*Blood serum metabolites (Means ± SEM) in dairy heifers fed on control and foreign probiotic yeast.*

| Parameters | Feeding regime | | |
|---|---|---|---|
| | Control[2] | IPY[3] | FPY[4] |
| **Urea (mg/100 ml)1** | | | |
| Before treatment[5] | 14.55 ± *0.57 | 14.18 ± 0.21 | 15.54 ± 0.32 |
| After treatment[6] | 14.18$^{a}$ ± 0.58 | 12.31$^{b}$ ± 0.22 | 13.68$^{ab}$ ± 0.90 |
| **Glucose (mg/100 ml)** | | | |
| Before treatment | 75.70 ± 1.24 | 73.99 ± 2.51 | 75.08 ± 2.30 |
| After treatment | 73.84$^{b}$ ± 0.71 | 77.42$^{a}$ ± 1.28 | 78.97$^{a}$ ± 0.54 |

*$^{a,b}$ Values on the same row with different superscripts differ significantly ($P < 0.05$).[1]n = 3 per treatment.*
*[2]Control feed without yeast.*
*[3]LAB-Probiotic feed compose of control feed supplemented with 3.13 × 10$^{07}$ cfu/g laboratory produces probiotic yeast (QAUSC03) at the rate of 8 g/day/animal.*
*[4]COM-Probiotic feed compose of control feed supplemented with 2.5×10 $^{07}$ cfu/g commercially probiotic yeast (Yac-Sac$^{1026}$) at the rate of 10g/day/animal.*
*[5]Before treatment (day 0).*
*[6]After treatment (day 120) * ± SEM = standard error of the mean.*

**Table 2.**
*Effect of indigenous Vs foreign probiotic yeast on blood parameters (Means ± SEM) in lactating dairy cattle.*

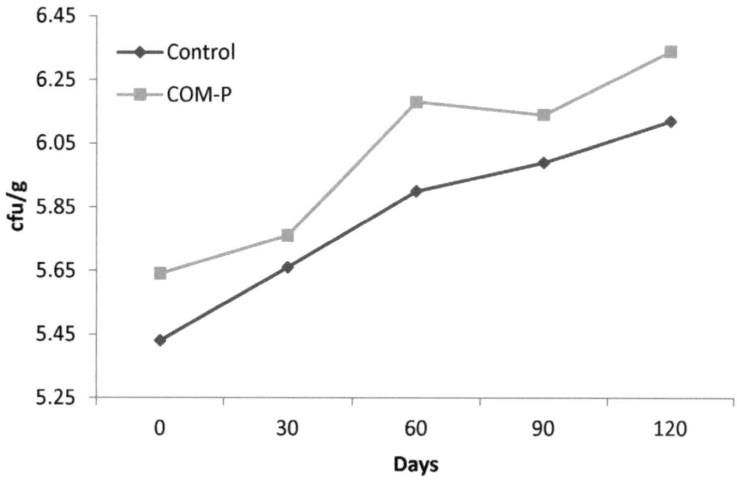


**Figure 7.**
*Total Lactococcus count (CFU/g) in the ruminal gut of dairy heifers fed on control feed (control, ♦; no yeast) or commercial probiotic feed (COM-P, ■; control feed plus commercial yeast) (n = 4).*

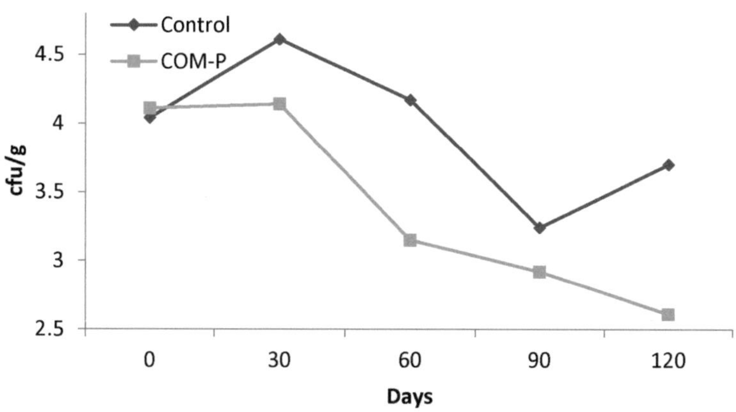


**Figure 8.**
*Total Enterococcus count (CFU/g) in the ruminal gut of dairy heifers fed on control feed (control, ♦; no yeast) or commercial probiotic feed (COM-P, ■; control feed plus commercial yeast) (n = 4).*

That means, we had a proof of positive impact of PFY on animal health. We had isolated the yeast from dairy animals fed on yeast. After careful assessment of the probiotic potential, we conducted another experiment to determine the impact of FPY Vs IPY on the health of lactating dairy cattle. Mix breed (*Sahiwal* and *Sahiwal×Jersey,* n = 9, with 4-5-liter milk per day) animal were selected for blood and fecal flora study. Animals were divided into three groups. Group 1 fed on 8 g IPY with 3.13 × $10^{07}$ CFU/g; group 2 fed on 10 g FPY with 2.5 ×1 $0^{07}$ CFU/g FPY, group 3 fed only control diet with no probiotic (Figure). After 90 days, results showed that the gut associated microbial flora and blood biochemical parameters were improved in the presence IPY as compare to the FPY (**Tables 1** and **2**).

We highlighted that improved animal health condition might be due to improved digestive enzymes produced from well propagated IPY. The VFAs have a capability to reduce the triglycerol and cholesterol in liver cells and might be change the animal lipid profile. Results of the ruminal gut microflora showed that the average, beneficial *Pediococcus* and *Weisella* species (CFU/g) counts increased while pathogenic E.coli species (CFU/g) counts decreased in IPY fed lactating cows than other groups which leads to improve GIT microbial balance (**Figures** 7 and **8**).

It can be concluded IPY improves the, gut health, and wellbeing of lactating dairy cattle in cost effective manner. IPY strain may adopt well in the cattle gut than FPY [80].

## 8. Conclusion

Ruminants of developing and developed countries have different types of gut microbiota due to their living standard, feeding type, their managemental style. Although from above discussion we have a clear understanding that the interlink between gut microbiota and fiber digestion plays a key role for obtaining maximum profit from dairy animals. Therefore, the PY must be target specific which give maximum outcomes in cost effective manners. For animals of specific geographical region, a unique and precise YP must be designed by isolating the local yeast strains from that population, only then maximum beneficial outputs can be obtained. The reason beings, compactivity of the local strains with normal microbiota of the rumen ecosystem (**Figure 9**).

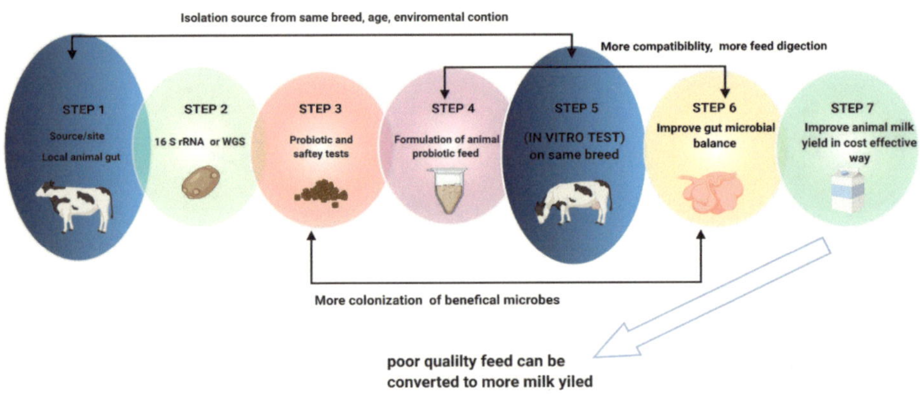


**Figure 9.**

*Target based Probiotic Preparation strategy: This figure showed probiotic preparation of by using the local animal GIT tract as preparation of local yeast probiotic. Interlinked factors involved in the application of probiotics in the ruminant's nutrition.*

## 9. Recommendations

The recommendations are outlined as follows;

- Pre-plane feed formulation for the manipulation of the rumen microbiota to digest the fibrous feed
- Identification of breed specific probiotic strains with same target.
- Whole genome sequencing of the probiotic strains as well as animal for maximum outputs
- Mode of action of the probiotic should studied well for understanding of the useful and useless probiotic.

## Author details

Shakira Ghazanfar
Functional Genomics and Bioinformatics, National Institute of Genomics and Agriculture Biotechnology (NIGAB), National Agricultural Research Centre, Islamabad, Pakistan

*Address all correspondence to: shakira_akmal@yahoo.com

## References

[1] Akin DE, Benner R. Degradation of polysaccharides and lignin by ruminal bacteria and fungi. 1988; Applied and Environmental Microbiology 54 1117-1125.

[2] Akin DE. Histological and Physical Factors Affecting Digestibility of Forages. 1989; Agronomy Journal 81 17-25.

[3] Alayande, Kazeem Adekunle, Olayinka Ayobami Aiyegoro, Thizwilondi Michael Nengwekhulu, Lebogang Katata-Seru, and Collins Njie Ateba. "Integrated genome-based probiotic relevance and safety evaluation of *Lactobacillus reuteri* PNW1." Plos one 15, no. 7 (2020): e0235873.

[4] Arakaki L, Stahringer R, Garrett J, Dehority B. The effects of feeding monensin and yeast culture, alone or in combination, on the concentration and generic composition of rumen protozoa in steers fed on low-quality pasture supplemented with increasing levels of concentrate. 2000; Animal Feed Science and Technology 84 121-127.

[5] Beev, G., P. Todorova, and S. Tchobanova. "Yeast cultures in ruminant nutrition." Bulgarian Journal of Agricultural Science 13 (2007): 357-374.

[6] Bonhomme A. Rumen ciliates: their metabolism and relationships with bacteria and their hosts. 1990; Animal Feed Science and Technology 30 203-266.

[7] Callaway E, Martin S. Effects of a *Saccharomyces cerevisiae* culture on ruminal bacteria that utilize lactate and digest cellulose. 1997; Journal of Dairy Science 80 2035-2044.

[8] Chaucheyras-Durand F, Chevaux E, Martin C, Forano E. Use of Yeast Probiotics in Ruminants: Effects and Mechanisms of Action on Rumen pH, Fibre Degradation, and Microbiota According to the Diet. 2012; Probiotic in Animals. InTech Open Book Publisher.

[9] Chaucheyras-Durand, F., and H. Durand. "Probiotics in animal nutrition and health." Beneficial microbes 1, no. 1 (2010): 3-9.

[10] Chaucheyras-Durand, F., N. D. Walker, and A. Bach. "Effects of active dry yeasts on the rumen microbial ecosystem: Past, present and future." Animal Feed Science and Technology 145, no. 1-4 (2008): 5-26.

[11] Chaucheyras-Durand, F., N. D. Walker, and A. Bach. "Effects of active dry yeasts on the rumen microbial ecosystem: Past, present and future." Animal Feed Science and Technology 145, no. 1-4 (2008): 5-26.

[12] Chaucheyras-Durand, Frédérique, Eric Chevaux, Cécile Martin, and Evelyne Forano. "Use of yeast probiotics in ruminants: Effects and mechanisms of action on rumen pH, fibre degradation, and microbiota according to the diet." Probiotic in animals (2012): 119-152.

[13] Chenoll, Empar, Inmaculada Moreno, María Sánchez, Iolanda Garcia-Grau, Ángela Silva, Marta González-Monfort, Salvador Genovés et al. "Selection of new probiotics for endometrial health." *Frontiers in cellular and infection microbiology* 9 (2019): 114.

[14] Cox, Faith, Peter H. Janssen, Gemma Henderson, Arjan Jonker, Wayne Young, and Siva Ganesh. "Rumen microbial community composition varies with diet and host, but a core microbiome is found across a wide geographical range." (2015).

[15] Crossland, Whitney Lynn, Aaron Bradley Norris, Luis Orlindo Tedeschi, and Todd Ryan Callaway. "Effects of active dry yeast on ruminal pH characteristics and energy partitioning

of finishing steers under thermoneutral or heat-stressed environment." Journal of animal science 96, no. 7 (2018): 2861-2876.

[16] Czerucka, D., T. Piche, and P. Rampal. "yeast as probiotics–Saccharomyces boulardii." Alimentary pharmacology & therapeutics 26, no. 6 (2007): 767-778.

[17] De Mulder, Thijs, Nico Peiren, Leen Vandaele, Tom Ruttink, Sam De Campeneere, Tom Van de Wiele, and Karen Goossens. "Impact of breed on the rumen microbial community composition and methane emission of Holstein Friesian and Belgian Blue heifers." Livestock Science 207 (2018): 38-44.

[18] Dolezal P, Dolezal J, Szwedziak K, Dvoracek J, Zeman L, Tukiendorf M, Havlicek Z. Use of Yeast Culture in the TMR of Dairy Holstein Cows. 2012; Iranian Journal of Applied Animal Science 2 51-56.

[19] Elghandour MM, Salem AZ, Castañeda JS, Camacho LM, Kholif AE, Chagoyán JC. Direct-fed microbes: A tool for improving the utilization of low quality roughages in ruminants. 2015; Journal of Integrative Agriculture 14 526-533

[20] Elghandour, M. M. Y., Tan, Z. L., Abu Hafsa, S. H., Adegbeye, M. J., Greiner, R., Ugbogu, E. A., ... & Salem, A. Z. M. (2020). Saccharomyces cerevisiae as a probiotic feed additive to non and pseudo-ruminant feeding: a review. Journal of applied microbiology, 128(3), 658-674.

[21] Elghandour, M. M. Y., Z. L. Tan, S. H. Abu Hafsa, M. J. Adegbeye, R. Greiner, E. A. Ugbogu, J. Cedillo Monroy, and A. Z. M. Salem. "Saccharomyces cerevisiae as a probiotic feed additive to non and pseudo-ruminant feeding: a review." Journal of applied microbiology 128, no. 3 (2020): 658-674.

[22] El-Ghani AA. Influence of diet supplementation with yeast culture (*Saccharomyces cerevisiae*) on performance of Zaraibi goats. 2004; Small ruminant research 52 223-229.

[23] Elliott, Christopher L., Joan E. Edwards, Toby J. Wilkinson, Gordon G. Allison, Kayleigh McCaffrey, Mark B. Scott, Pauline Rees-Stevens, Alison H. Kingston-Smith, and Sharon A. Huws. "Using 'Omic approaches to compare temporal bacterial colonization of Lolium perenne, Lotus corniculatus, and Trifolium pratense in the rumen." Frontiers in microbiology 9 (2018): 2184.

[24] Elliott, Christopher L., Joan E. Edwards, Toby J. Wilkinson, Gordon G. Allison, Kayleigh McCaffrey, Mark B. Scott, Pauline Rees-Stevens, Alison H. Kingston-Smith, and Sharon A. Huws. "Using 'Omic approaches to compare temporal bacterial colonization of Lolium perenne, Lotus corniculatus, and Trifolium pratense in the rumen." Frontiers in microbiology 9 (2018): 2184.

[25] Fernández-Pacheco, Pilar, Carolina Cueva, María Arévalo-Villena, M. Victoria Moreno-Arribas, and Ana Briones Pérez. "Saccharomyces cerevisiae and Hanseniaspora osmophila strains as yeast active cultures for potential probiotic applications." *Food & function* 10, no. 8 (2019): 4924-4931.

[26] Ferraretto LF, Shaver RD, Bertics SJ. Effect of dietary supplementation with live-cell yeast at two dosages on lactation performance, ruminal fermentation, and total-tract nutrient digestibility in dairy cows. 2012; Journal of Dairy Science 95 4017-4028.

[27] Fuller R. Probiotics in man and animals. 1989;The Journal of Applied Bacteriology 66 365-378.

[28] Garcia-Mazcorro, J. F., S. L. Ishaq, M. V. Rodriguez-Herrera, C. A. Garcia-Hernandez, J. R. Kawas, and T. G. Nagaraja. Review"Are there indigenous

Saccharomyces in the digestive tract of livestock animal species? Implications for health, nutrition and productivity traits." animal 14, no. 1 (2020): 22-30.

[29] Ghazanfar S. Study on the effects of dietary supplementation of Saccharomyces cerevisiae on performance of dairy cattle and heifers. 2016; PhD Thesis. Quaid-i-Azam University, Islamabad. Pakistan

[30] Ghazanfar S. Study on the effects of dietry supplmentation of *Saccharomyvces cerevisiae* on performance of dairy cattle and heifers. 2016; PhD Thesis. Quaid-i-Azam University, Islamabad. Pakistan

[31] Ghazanfar, Shakira, Nauman Khalid, Iftikhar Ahmed, and Muhammad Imran. "Probiotic yeast: mode of action and its effects on ruminant nutrition." Yeast—Industrial Applications, IntechOpen (2017): 179-202.

[32] Ghazanfar, Shakira, Nauman Khalid, Iftikhar Ahmed, and Muhammad Imran. "Probiotic yeast: mode of action and its effects on ruminant nutrition." Yeast—Industrial Applications, IntechOpen (2017): 179-202.

[33] Giang, Hoang Huong, Tran Quoc Viet, Brian Ogle, and Jan Erik Lindberg. "Effects of supplementation of probiotics on the performance, nutrient digestibility and faecal microflora in growing-finishing pigs." Asian-Australasian Journal of Animal Sciences 24, no. 5 (2011): 655-661.

[34] Gijzen HJ, Lubberding HJ, Gerhardus MJT, Vogels GD. Contribution of rumen protozoa to fibre degradation and cellulase activity in vitro. 1988; FEMS Microbiology Letters 53 35-43.

[35] He, Z. X., B. Ferlisi, E. Eckert, H. E. Brown, A. Aguilar, and M. A. Steele. "Supplementing a yeast probiotic to pre-weaning Holstein calves: Feed intake, growth and fecal biomarkers of gut health." Animal Feed Science and Technology 226 (2017): 81-87.

[36] Henderson, Gemma, Faith Cox, Siva Ganesh, Arjan Jonker, Wayne Young, Leticia Abecia, Erika Angarita et al. "Rumen microbial community composition varies with diet and host, but a core microbiome is found across a wide geographical range." Scientific reports 5 (2015): 14567.

[37] Hillal, Hany, Gamal El-Sayaad, and Mohamed Abdella. "Effect of growth promoters (probiotics) supplementation on performance, rumen activity and some blood constituents in growing lambs." Archives Animal Breeding 54, no. 6 (2011): 607-617.

[38] Hungate, R. E. "The rumen microbial ecosystem." Annual Review of Ecology and Systematics 6, no. 1 (1975): 39-66.

[39] Jackson, Scott A., Jean L. Schoeni, Christina Vegge, Marco Pane, Buffy Stahl, Michael Bradley, Virginia S. Goldman, Pierre Burguière, John B. Atwater, and Mary Ellen Sanders. "Improving end-user trust in the quality of commercial probiotic products." Frontiers in microbiology 10 (2019): 739.

[40] Joblin K. Physical disruption of plant fibre by rumen fungi of the Sphaeromonas group. The Roles of Protozoa and Fungi in Ruminant Digestion. 1989; Penambul Armidale NSW 259-260.

[41] Kamra, Devki Nandan. "Rumen microbial ecosystem." Current science (2005): 124-135.

[42] Khatri, Indu, Rajul Tomar, K. Ganesan, G. S. Prasad, and Srikrishna Subramanian. "Complete genome sequence and comparative genomics of the probiotic yeast Saccharomyces boulardii." Scientific reports 7, no. 1 (2017): 1-12.

[43] Konsue, Wilasinee, Tida Dethoup, and Savitree Limtong. "Biological control of fruit rot and anthracnose of postharvest mango by antagonistic yeasts from economic crops leaves." Microorganisms 8, no. 3 (2020): 317.

[44] Krause, D. O., T. G. Nagaraja, A. D. G. Wright, and T. R. Callaway. "Board-invited review: rumen microbiology: leading the way in microbial ecology." Journal of animal science 91, no. 1 (2013): 331-341.

[45] Lesmeister, K. E., Arlyn Judson Heinrichs, and M. T. Gabler. "Effects of supplemental yeast (Saccharomyces cerevisiae) culture on rumen development, growth characteristics, and blood parameters in neonatal dairy calves." Journal of dairy science 87, no. 6 (2004): 1832-1839.

[46] Li, Fuyong, Changxi Li, Yanhong Chen, Junhong Liu, Chunyan Zhang, Barry Irving, Carolyn Fitzsimmons, and Graham Plastow. "Host genetics influence the rumen microbiota and heritable rumen microbial features associate with feed efficiency in cattle." *Microbiome* 7, no. 1 (2019): 92.

[47] MA, Song-cheng, Jing CHEN, and Hua-ming MAO. "Rumen Microbial Ecosystem [J]." China Animal Husbandry & Veterinary Medicine 1 (2007).

[48] Martinez-Fernandez, Gonzalo, Stuart E. Denman, Chunlei Yang, Jane Cheung, Makoto Mitsumori, and Christopher S. McSweeney. "Methane inhibition alters the microbial community, hydrogen flow, and fermentation response in the rumen of cattle." Frontiers in Microbiology 7 (2016): 1122.

[49] Mestecky J, Russell M. Passive and active protection against disorders of the gut. 1998; Veterinary Quarterly 20 83-87.

[50] Morgavi, D. P., Evelyne Forano, Cécile Martin, and C. Jamie Newbold. "Microbial ecosystem and methanogenesis in ruminants." Animal: an international journal of animal bioscience 4, no. 7 (2010): 1024.

[51] Morgavi, Diego, William Kelly, Peter Janssen, and Graeme Attwood. "Rumen microbial (meta) genomics and its application to ruminant production." Animal 7, no. Suppl. 1 (2013): 184-201.

[52] Musa H, Wu S, Zhu C, Seri H, Zhu G. The potential benefits of probiotics in animal production and health. 2009; Journal of Animal and veterinary Advances 8 313-321.

[53] Newbold C, Wallace R, Chen X, McIntosh F. Different strains of *Saccharomyces cerevisiae* differ in their effects on ruminal bacterial numbers in vitro and in sheep. 1995; Journal of Animal Science 73 1811-1818.

[54] Newbold, C. J. "Probiotics for ruminants." (1996).

[55] Newbold, C. J., and Eva Ramos-Morales. "Ruminal microbiome and microbial metabolome: effects of diet and ruminant host." animal 14, no. S1 (2020): s78-s86.

[56] Nielsen, Jens. "Yeast systems biology: model organism and cell factory." Biotechnology journal 14, no. 9 (2019): 1800421.

[57] Nurmi E, Rantala M (1973) New aspects of *Salmonella* infection in broiler production. Nature 241 210-211.

[58] Paul SS, Kamra DN, Sastry VR, Sahu NP, Kumar A. Effect of phenolic monomers on biomass and hydrolytic enzyme activities of an anaerobic fungus isolated from wild nil gai (Baselophus tragocamelus). 2003; Letters in Applied Microbiology 36 377-381.

[59] Poppy GD, Rabiee AR, Lean IJ, Sanchez WK, Dorton KL. A meta-analysis of the effects of feeding

yeast culture produced by anaerobic fermentation of Saccharomyces cerevisiae on milk production of lactating dairy cows. 2012; Journal of Dairy Science 95 6027-6041.

[60] Poppy GD, Rabiee AR, Lean IJ, Sanchez WK, Dorton KL. A meta-analysis of the effects of feeding yeast culture produced by anaerobic fermentation of *Saccharomyces cerevisiae* on milk production of lactating dairy cows. 2012; Journal of Dairy Science 95 6027-6041

[61] Robinson PH, Erasmus LJ. Effects of analyzable diet components on responses of lactating dairy cows to Saccharomyces cerevisiae based yeast products: A systematic review of the literature. 2009; Animal Feed Science and Technology 149 185-198.

[62] Robinson PH, Erasmus LJ. Effects of analyzable diet components on responses of lactating dairy cows to *Saccharomyces cerevisiae* based yeast products: A systematic review of the literature. 2009; Animal Feed Science and Technology 149 185-198.

[63] Russell, James B., and Robert B. Hespell. "Microbial rumen fermentation." Journal of Dairy Science 64, no. 6 (1981): 1153-1169.

[64] Santra A, Karim S. Rumen manipulation to improve animal productivity. 2003; Asian Australasian Journal of Animal Sciences 16 748-763.

[65] Shakira G, Atiya A Ahmed.I. Effects of Dietary Supplementation of Yeast (Saccharomyces cerevisiae) Culture on Growth Performance, Blood Parameters, Nutrient Digestibility and Fecal Flora of Dairy Heifers. 2015; The Journal of Animal and Plant Science 25 53-59.

[66] Srinivasan, Prashanth, and Christina D. Smolke. "Biosynthesis of medicinal tropane alkaloids in yeast." Nature 585, no. 7826 (2020): 614-619.

[67] Steele, Michael A., Greg B. Penner, and Frédérique Chaucheyras-Durand. "Development and physiology of the rumen and the lower gut: Targets for improving gut health." Journal of dairy science 99, no. 6 (2016): 4955-4966.

[68] Swanson, Kelly S., Glenn R. Gibson, Robert Hutkins, Raylene A. Reimer, Gregor Reid, Kristin Verbeke, Karen P. Scott et al. "The International Scientific Association for Probiotics and Prebiotics (ISAPP) consensus statement on the definition and scope of synbiotics." Nature Reviews Gastroenterology & Hepatology 17, no. 11 (2020): 687-701.

[69] Tapio, Ilma, Daniel Fischer, Lucia Blasco, Miika Tapio, R. John Wallace, Ali R. Bayat, Laura Ventto et al. "Taxon abundance, diversity, co-occurrence and network analysis of the ruminal microbiota in response to dietary changes in dairy cows." *PloS one* 12, no. 7 (2017): e0180260.

[70] Tripathi VK, Sehgal JP, Puniya AK, Singh K. Hydrolytic activities of anaerobic fungi from wild blue bull (Boselaphus tragocamelus). 2007; Anaerobe 13: 36-39.

[71] Tripathi VK, Sehgal JP, Puniya AK, Singh K. Hydrolytic activities of anaerobic fungi from wild blue bull (Boselaphus tragocamelus). 2007; Anaerobe 13: 36-39.

[72] Ushida K, Jouany JP. Effect of defaunation on fibre digestion in sheep given two isonitrogenous diets. 1990; Animal Feed Science and Technology 29 153-158.

[73] Wallace, R. J. "Ruminal microbiology, biotechnology, and ruminant nutrition: progress and problems." Journal of Animal Science 72, no. 11 (1994): 2992-3003.

[74] Wallace, R. John, and C. James Newbold. "Probiotics for ruminants." In Probiotics, pp. 317-353. Springer, Dordrecht, 1992.

[75] Wallace, R. John. "Rumen microbiology, biotechnology and ruminant nutrition: the application of research findings to a complex microbial ecosystem." *FEMS microbiology letters* 100, no. 1-3 (1992): 529-534.

[76] Wiles, Travis J., Matthew Jemielita, Ryan P. Baker, Brandon H. Schlomann, Savannah L. Logan, Julia Ganz, Ellie Melancon, Judith S. Eisen, Karen Guillemin, and Raghuveer Parthasarathy. "Host gut motility promotes competitive exclusion within a model intestinal microbiota." *PLoS biology* 14, no. 7 (2016): e1002517.

[77] Williams AG, Orpin CG. Polysaccharide-degrading enzymes formed by three species of anaerobic rumen fungi grown on a range of carbohydrate substrates.1987; Canadian Journal of Microbiology 33 418-426.

[78] Williams AG, Orpin CG. Polysaccharide-degrading enzymes formed by three species of anaerobic rumen fungi grown on a range of carbohydrate substrates.1987; Canadian Journal of Microbiology 33 418-426.

[79] Yang CM, Varga GA. The effects of continuous ruminal dosing with dioctyl sodium sulphosuccinate on ruminal and metabolic characteristics of lactating Holstein cows. 1993; British Journal of Nutrition 69 397-408.

[80] Youngblut, Nicholas D., Georg H. Reischer, William Walters, Nathalie Schuster, Chris Walzer, Gabrielle Stalder, Ruth E. Ley, and Andreas H. Farnleitner. "Host diet and evolutionary history explain different aspects of gut microbiome diversity among vertebrate clades." Nature communications 10, no. 1 (2019): 1-15.

Section 2

# Medical and Pharmaceutical Applications

Chapter 4

# Saccharomyces: Is a Necessary Organism or a Biological Warrior?

*Nilay Seyidoglu and Cenk Aydin*

## Abstract

Saccharomyces is a eukaryotic organism that possesses approximately 6,000 known genes since 1996. It has long been used for food, bakeries, drinks, and therapeutics due to its many ingredients and its role in several mechanisms. Saccharomyces can be used as an experimental organism for medicinal products in the pharmaceutical industry. Particularly in public health, the use of Saccharomyces in the production of vaccines is remarkable. It has been alleviated that this yeast helps clarify the function of individual proteins in pathogenic viruses. To clarify virus life and host interactions, virus replication systems in Saccharomyces were interested in scientists. The new antiviral strategies with yeasts suggest the biological mechanism of a pathogen virus. Due to the variety of diseases and current epidemic conditions, these organisms play an essential role in prevention and treatment. This chapter will try to update Saccharomyces' scientific discoveries with the most recent and up-to-date literature.

**Keywords:** Saccharomyces, pandemic diseases, experimental organisms, public health, antiviral strategies

## 1. Introduction

Besides poor treatment and vaccination programs, a healthy immune system and antioxidant mechanism are the essential defenders considering the current viral diseases. The viral diseases hosted in a body has several impacts on organs and systems. Also, long-term drug use or vaccination programs can cause some acute side effects on the body, such as gut microbiota, immunity, lung tissue, etc. Therefore, probiotics, prebiotics, vitamins, natural antioxidants have been generally recommended over the years. Probiotics named live microorganisms have beneficial effects on systems, and they have been used successfully. Prebiotics are non-digestible foods that stimulate intestinal tissue growth and modulate immunity. Vitamins, minerals, and natural antioxidants have been used to enhance immune activity and health in viral diseases. It can be said that all these supplements are essential for adequate homeostasis.

Today, evaluate the most effective, economical, and safe vaccines is a significant challenge. Thus, some crucial organisms have been interested in vaccine production as well as nutrition. Among the different vaccination process, yeasts have a broad interest in the scientific area (**Figure 1**). These unicellular and saprotrophic fungi have been used as a biological model. They have also been accepted as critical models for experiments due to their cellular structure, components, and rapid growth. Yeast also can be cultured easily and manipulated genetically. These features showed that

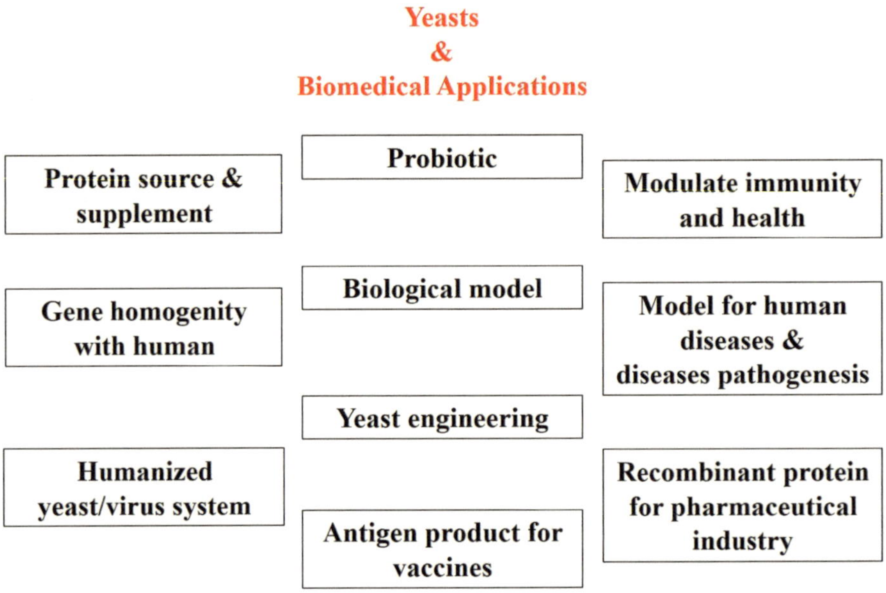


**Figure 1.**
*Biomedical applications of yeasts.*

yeasts are beneficial to identify the cellular mechanism of virus and vaccine programs safely [1, 2].

The yeast Saccharomyces, the essential eukaryotic organism, have been used as a biological model. Nevertheless, there is a notable gene homology in this yeast with human genes. In this chapter, we try to identify the Saccharomyces yeast as a useful model for biological experiments and observe the importance of viruses, viral diseases, and vaccines.

## 2. Saccharomyces

*Saccharomyces cerevisiae* is a model organism extensively used to investigate the biology of eukaryotic cells. It is widely used as a cell factory for producing pharmaceuticals, chemicals, and biofuels [3].

Saccharomyces, which is a genus belonging to the Saccharomyces fungus kingdom, includes many yeast species. The name of Saccharomyces is derived from the Latin words saccharo- (sugar) and - Greek mikes (mushrooms). These yeasts were initially suggested in 1680, and named Saccharomyces in 1837. A successful systemic concept on these higher eukaryotes was designed by Mayr [4]. Yeasts' cultured forms have been used for thousands of years due to rapid reproduction and essential components. Typical features of Saccharomyces are the usage of nitrate and ability for the fermentation of carbohydrates. Saccharomyces have an excellent capacity for ethanol production, and they are suitable yeasts for large-scale fermentation [5]. These important yeasts can be used for the food industry to produce several foods such as bread, beer, wine, distilled spirits, and industrial alcohols. The most knows are *S. cerevisiae, S. boullardii, S. pombe, S. pastorianus*, and *S. paradoxus*, mostly used for food and treatments. Nevertheless, these yeasts have a small nucleus and central vacuole and have glucan and mannoproteins on their cell walls. Saccharomyces include a single linear double-stranded DNA, ribosomal proteins, and non-ribosomal molecules, like other eukaryotes. It was suggested that their genetic structure is beneficial for the model organism, especially *S. cerevisiae* [6].

*S. cerevisiae* a single celled organism that is used as a model organism. These yeasts have been studied to understand the concept of cell cycle regulation, DNA repair, and other cellular mechanisms. It was also reviewed that a model to identify the mutations in the cell cycle in cancer and some diseases, especially neurodegenerative diseases [7]. However, a form of *S. cerevisiae* called *S. boulardii* had been observed in clinical trials for treatment such as inflammation and diarrhea. Mc Farland and Bernasconi reported that *S. boulardii* is a wild type of Saccharomyces, a pharmaceutical agent [8]. The action of *S. boulardii* has been described by releasing trypsin-like protease, which inhibits the toxins in inflammations [9].

*Schizosaccharomyces pombe* is a fission yeast that was isolated in 1893 by Paul Lindner from East African millet beer. It is a model organism for eukaryotic cell biology and molecular biology as well as *S. boulardii* and *S. cerevisiae*. In 1590, Mitchison was firstly studied with this yeast in an experimental organism. Eser et al. reported that it could be used to treat diabetes and other diseases [10]. This fission yeast has been studied for eukaryotic RNA metabolism due to its gene expression.

*S. bayanus (S. eubayanus), S. paradoxus*, and *S. pastorianus* have similar genome size with *S. cerevisiae*. They all have been studied for DNA reassociation studies [11]. *S. pastorianus* is a lager yeast, an interspecific hybrid between *S. cerevisiae* and another *S. bayanus (S. eubayanus)* [12]. It is also a synonym of *S. carlsbergensis* and closely related to genus *S. cerevisiae*. Another wild type of yeast, *S. paradoxus*, can be isolated from nature, especially tree exudates or oils. It is an essential type of yeast for genetic and genomic studies. A yeast named *S. bayanus (S. eubayanus)*, which was isolated from the tree, is related to *S. cerevisiae* and *S. pastorianus* [13]. *S. bayanus* has been used for genomic studies, expression patterns, and nucleosomes profiles [14–16].

Saccharomyces yeasts focus on the dietary field as a probiotic and the process of treating the disease. Belong the probiotic action; these yeasts have several vital roles on mechanisms such as bacterial adhesion, enhancement of immune cells and responses, modulation of the signaling pathways of the host, and improvement of the strengthening of enterocytes [17]. Nevertheless, Saccharomyces are used as model organisms in biological studies, particularly chemicals and pharmaceuticals.

## 3. Experimental organism for pharmaceutical industry

Over the last fifty years, remarkable progress in our ability to produce advanced drugs has improved people's health and longevity. Pharmaceutical proteins are one of the fastest-growing groups of medicines and are currently critical to treating many diseases [18].

Proteins have a catalyzer role in several metabolic reactions as well as an essential responsibility for cellular mechanisms. There are unique systems that can be used to produce proteins for the pharmaceutical industry from a single cell to multiple organisms, including eukaryotes, especially yeasts. Dozens of pharmaceutical proteins, including insulin, vaccines, and blood factors, produced by *S. cerevisiae*, have been commercialized. It was reviewed that yeasts are essential for biological activities, mainly producing the purified product due to its cost-effective, fast production like bacteria and high density of cell cultures [19]. In recent years, indeed, as a model organism, yeasts have been provided to identify the pathogenesis and role for diseases, especially *S. cerevisiae* and *S. pombe.*

The yeast Saccharomyces has been accepted as the significant organism for several metabolisms such as cell cycle, biogenesis, protein folding, genetic manipulation, recombination, etc. [20]. *S. cerevisiae* is a unicellular microbial organism that grows fast, tolerances to chemicals, and cultured easily. It was reported that this yeast could discover the process of diseases because of the conservation of molecular interactions from yeast to humans [21, 22]. On the other hand, *S. cerevisiae* can be an essential organism for recombinant protein production for pharmacy. It has full cellular organelles and membrane compartments that produce many eukaryotic proteins, including humans' [23]. Initially, the essential biopharmaceuticals insulin and its analogs have been produced by *S. cerevisiae*. Researchers have reported other important biopharmaceuticals such as the human serum albumin, hepatitis vaccines, and virus-like particles for vaccination (**Table 1**). Also, several medicines have been produced with *S. cerevisiae* until 2012 reported by the European Medicines Agency [18]. Furthermore, current studies showed that metabolic engineering pathways and optimization procedures of *S. cerevisiae* are essential for producing recombinant proteins for pharmaceuticals and biomedical areas [18, 19]. *S. cerevisiae* carries out human-like glycoprotein that is efficient for producing recombinant proteins. Protein secretion of *S. cerevisiae* is complex processing that follows as transcription, translation, translocation, post-translational modifications, folding, peptide cleavage, glycosylation, sorting, and secretion. This important organism enables genetic modifications. It was reported that the first eukaryotic organism sequenced DNA in *S. cerevisiae* [41]. Due to the protein misfolding and aggregation, *S. cerevisiae* has been used as a model organism.

Nevertheless, *S. pombe* has been accepted as a model organism together with *S. cerevisiae*. This fission yeast is used as a successful host. It was reviewed that *S. pombe* and generated strains have significant facilitation for producing drug glucuronides [42, 43]. The classical yeast genetics approaches can be described for *S. pombe*. It has been accepted as the most ancient yeast molecule. However, *S. pombe* has been more advanced evolutionarily than other yeasts. *S. pombe* has become a model organism until 2002 [44, 45].

Recombinant proteins are recognized as an important part of the drug industry. Among these proteins, Saccharomyces has greater attention than others due to their eukaryotic properties, easy genetic manipulation, and capable of modifications. *S. cerevisiae* emerges as the most common host to express heterologous genes and therapeutic proteins [46]. This organism may provide a simple background for isotype expressions, and thereby drug metabolism studies can be easily associated with genome screens, underlying toxicity, and encoded genomes.

| Bioparhamaceutical products | Category | References |
|---|---|---|
| Human serum albumin | Blood factors | Payne et al. [24] |
| Recombinant proteins | Protein | Huang et al. [18], Ferrer-Millares et al. [19], Ma et al. [25], Cino [26] |
| Insulin | Hormone | Martinez et al. [27] |
| Glucagon | Hormone | Egel-Mitani et al. [28] |
| Human parathyroid hormone | Hormone | Song et al. [29] |
| Purified protein for vaccines | Protein | Hadiji-Abbes et al. [30], Zhang et al. [31], King et al. [32], Kaslow and Shiloach [33], Fazlalipour et al. [34]. |
| Virus like particles | Protein | Jacobs et al. [35], Kim et al. [36], Kim et al. [37]. |
| Gene expression systems | Gene | Malak et al. [38], van Ooyen et al. [39], Vierira Gomes et al. [40]. |

**Table 1.**
*Examples of bioparhamaceutical products of Saccharomyces.*

## 4. Antiviral strategies

While the vaccines currently available have proven invaluable in the fight against infectious diseases and eradicating viruses, there are many drawbacks to the current vaccine preparation or application regimen despite these successes. Certain limitations of conventional vaccines require multiple adjuvants and injections to induce a necessary or optimal immune response. Another reason is the constant increase in the number of post-vaccination allergic reactions or hypersensitivities in a specific group of people [47, 48].

Today, there are several critical viral diseases such as human hepatitis B and C, immunodeficiency virus (HIV), severe acute respiratory syndrome coronavirus (SARS), coronavirus-disease 2019 (COVID19), etc. Due to the inadequacy of treatment options for these infections, new antiviral strategies and model organisms, particularly yeast, were of interest to the researchers.

Yeasts have a delivery system for nucleic acids, and thus they can be an alternative for virus description. Besides, a humanized yeast system was identified for yeast/virus systems to study diseases [49]. Yeasts are used for subunit vaccine formulations with producing antigens against viruses. It was reviewed that yeast can be used for vaccine development in such strategies; whole recombinant yeast, virus-like particles, yeast display, and purified protein immunogens [50]. Among yeasts, *S. cerevisiae* has been accepted as a versatile model organism for viruses' research, from the wire of public health to vaccine production.

Rosenfeld and Racaniello [51] reported that hepatitis C virus (HCV) was demonstrated in *S. cerevisiae*, and all proteins for the virus were encoded. Another study reported that *S. cerevisiae* could safely express the hepatitis B surface antigen in prophylactic vaccines [52]. Researchers observed that yeast could help clarify the function of viruses' proteins with dissection of RNA viruses' life cycle [53, 54]. Nevertheless, several protein immunogens can be purified from Saccharomyces. These immunogen proteins derived from yeasts are associated with virus-like particles. Virus-like particles can provide an alternative for viruses, and FDA approved this vaccine for hepatitis B and papillomavirus [55]. Also, the circumsporozoite protein derived from *S. cerevisae* is an immunodominant antibody of malaria. This preparation increased the antibodies and thereby neutralized the sporozoites [56]. Due to the yeast membrane permeability, *S. cerevisiae* enables entry to the chemical compounds and provides virus-host interactions. Some researchers showed that beta-glucan of the yeast cell wall could provide the immune response that important for vaccine development [57].

All things considered, the yeast-based carrier system can be a potential model to develop the vaccine insights of virus-host interactions. The yeast strategies can improve the recognition of pathogen antigens peptides, activate the immune response, and also modulate the yeast-based vaccines. Researchers for further pioneering findings have still endured the studies.

## 5. Future perspectives

There have been many illnesses that have not been controlled by vaccination and new ones as well. Mutation, genetic exchange, environmental and interspecific transference, or human contact are the most emerging diseases. However, new scientific technologies, model organisms and a number of researchers have proven beneficial to vaccination strategies. In this respect, it is possible to observe yeasts for the upcoming vaccines for several diseases.

Yeast engineered to the virus has been accepted as an ideal therapeutic approach. This vaccine's strategy is improving humoral immunity due to the ability of yeast to the generation of immune responses.

There is a numerous increasing study to obtain the vaccine strategy of yeasts. Studies in yeast proteins and cell wall components, including beta-glucan, may become more critical for vaccine strategies under different phases of clinical trials on animals or humans. According to the essential features of yeast, the yeast-based vaccine strategy is being necessary for vaccine development. It has foreseen that diversity of yeast strains will improve in the future.

## 6. Conclusion

The yeast system provides invaluable antiviral strategies. Significant studies have been conducted on yeast progression in the identification of viral diseases and antiviral strategies. Based on a better understanding of yeast protein and viruses, the search for new vaccines and medications for viral or pandemic diseases is safer and more effective. However, experiments with animal models and human cells are still underway in many types of yeast. Knowledge of these new biological systems and technologies, models, and organisms will open up new science avenues.

## Conflict of interest

The authors declare no conflict of interest.

## Acronyms and abbreviations

| | |
|---|---|
| HIV | Human immunodeficiency virus |
| SARS | Severe Acute Respiratory Syndrome |
| COVID-2019 | Coronavirus-disease 2019 |
| FDA | U.S. Food and Drug Administration |
| DNA | Deoxyribonucleic acid |
| RNA | Ribonucleic acid |
| *S. cerevisiae* | *Saccharomyces cerevisiae* |
| *S. boullardii* | *Saccharomyces boullardii* |
| *S. carlsbergensis* | *Saccharomyces carlsbergensis* |
| *S. bayanus* | *Saccharomyces bayanus* |
| *S. pastorianus* | *Saccharomyces pastorianus* |
| *S. paradoxus* | *Saccharomyces paradoxus* |
| *S. ebayanus* | *Saccharomyces ebayanus* |
| *S. pombe* | *Schizosaccharomyces pombe* |

## Appendices and nomenclature

| | |
|---|---|
| Yeast | The most important eukaryote; Saccharomyces. |
| Single celled organism | *Saccharomyces cerevisiae* |
| Nucleosomes | DNA, RNA |
| Biopharmaceuticals | insulin and its analogs |
| Eukaryotes | The organisms whose cells have a nucleus enclosed within a nuclear envelope |

## Author details

Nilay Seyidoglu[1*] and Cenk Aydin[2]

1 Department of Physiology, Faculty of Veterinary Medicine, Tekirdag Namik Kemal University, Tekirdag, Turkey

2 Department of Physiology, Faculty of Veterinary Medicine, Bursa Uludag University, Bursa, Turkey

*Address all correspondence to: nseyidoglu@nku.edu.tr

## References

[1] Wickner RB, Fujimura T, Esteban R. Viruses and prions of *Saccharomyces cerevisiae*. Adv Virus Res. 2013;86:1-36. doi:10.1016/B978-0-12-394315-6.00001-5

[2] Walch B, Breinig T, Schmitt MJ, Breinig F. Delivery of functional DNA and messenger RNA to mammalian phagocytic cells by recombinant yeast. Gene Therapy.2012;19:237-245. DOI: 10.1038/gt.2011.121

[3] Nielsen J, Keasling JD. Engineering cellular metabolism. Cell. 2016;164(6):1185-1197. DOI: 10.1016/j.cell.2016.02.004

[4] Mayr E. Systematics and the Origin of Species. New York: Columbia University Press; 1942.

[5] Zhao XQ, Bai FW. Mechanisms of yeast stress tolerance and its manipulation for efficient fuel ethanol production. J Biotechnol. 2009;144:23-30. DOI: 10.1016/j.jbiotec.2009.05.001.

[6] Del Giudice L, Massardo DR, Pontieri P, Wolf K. Interaction between yeast mitochondrial and nuclear genomes: Null alleles of RTG genes affect resistance to the alkaloid lycorine in rho(0) petites of *Saccharomyces cerevisiae*. Gene. 2005;354:9-14. DOI: 10.1016/j.gene.2005.03.020

[7] Galao RP, Scheller N, Alves-Rodrigues I, Breinig T, Meyerhans A, Diez J. *Saccharomyces cerevisiae*: a versatile eukaryotic system in virology. Microb Cell Fact. 2007;6:32. doi: 10.1186/1475-2859-6-32.

[8] Mcfarland LV, Bernasconi P. Saccharomyces boulardii: a review of an innovative biotherapeutic agent. Microbial Ecology in Health and Disease. 1993;6:157-171. DOI: 10.3109/08910609309141323

[9] Pothoulakis C, Kelly CP, Joshi MA, Gao N, O'Keane CJ, Castagliuolo I, Lamont JT. *Saccharomyces boulardii* inhibits Clostridium difficile toxin A binding and enterotoxicity in rat ileum. Gastroenterology. 1993;104:1108-1115. DOI: 10.1016/0016-5085(93)90280-p

[10] Eser P, Wachutka L, Maier KC, Demel C, Boroni M, Iyer S, Cramer P, Gagneur J. Determinants of RNA metabolism in the *Schizosaccharomyces pombe* genome. Molecular Systems Biology. 2016;12(2):857. DOI: 10.15252/msb.20156526

[11] Stewart GG. Saccharomyces, Introduction. In: Batt CA, Tortorello ML, editors. Encyclopedia of Food Microbiology. 2nd edition. Amsterdam: Elsevier; 2014. p.297-301.

[12] Boynton PJ, Greig D. The ecology and evolution of non-domesticated Saccharomyces species. Yeast. 2014: 31(12):449-462. DOI: 10.1002/yea.3040

[13] Libkind D, Hittinger CT, Vale'rio E, Gonçalves C, Dover J, Johnston M, Gonçalves P, Sampaio JP. Microbe domestication and the identification of the wild genetic stock of lager-brewing yeast. Proc Natl Acad Sci USA. 2011;108:14539-14544

[14] Guan Y, Dunham M, Caudy A, Troyanskaya O. Systematic planning of genome-scale experiments in poorly studied species. PLOS Computational Biology. 2010;6(3):e1000698. DOI: 10.1371/journal.pcbi.1000698

[15] Guan Y, Yao V, Tsui K, Gebbia M, Dunham MJ, Nislow C, Troyanskaya OG. Nucleosome-coupled expression differences in closely-related species. BMC Genomics. 2011;12:466. DOI: 10.1186/1471-2164-12-466. PMC 3209474. PMID 21942931

[16] Guan Y, Dunham MJ, Troyanskaya OG, Caudy AA. Comparative gene expression between two yeast species. BMC Genomics. 2013;14:33. doi:10.1186/1471-2164-14-33.

[17] Chen X, Kelly CP. Saccharomyces. In Versalovic J, Wilson M, editors. Therapeutic microbiology. Washington: ASM Press; 2008. p. 51-60 DOI: 10.1128/9781555815462.ch5

[18] Huang M, Bao J, Nielsen J. Biopharmaceutical protein production by *Saccharomyces cerevisiae*: current state and future prospects. Pharmaceutical Bioprocessing. 2014;2(2):167-182. DOI: 10.4155/pbp.14.8

[19] Ferrer-Miralles N, Domingo-Espin J, Corchero JL, Vazquez E, Villaverde A. Microbial factories for recombinant pharmaceuticals. Microb Cell Fact. 2009;8:17. DOI: 10.1186/1475-2859-8-17

[20] Pereira C, Countinho I, Soares J, Bessa C, Leaõ M, Saraiva L. New insights into cancer-related proteins provided by theyeast model. FEBS Journal. 2012;279:697-712. DOI: 10.1111/j.1742-4658.2012.08477.x

[21] Botstein D, Fink GR. Yeast: an experimental organism for modern biology. Science. 1988;240:1439-1443. DOI: 10.1126/science.3287619

[22] Smith MG, Snyder M. Yeast as a model for human disease. In: Curr Protoc Hum Genet. 2006. Chapter 15:Unit 15.6. DOI: 10.1002/0471142905.hg1506s48

[23] Nielsen J. Production of biopharmaceutical proteins by yeast: advances through metabolic engineering. Bioengineered. 2013;4(4):207-211. DOI: 10.4161/bioe.22856

[24] Payne T, Finnis C, Evans LR, Mead DJ, Avery SV, Archer DB, Sleep D. Modulation of chaperone gene expression in mutagenized *Saccharomyces cerevisiae* strains developed for recombinant human albumin production results in increased production of multiple heterologous proteins. Appl Environ Microbiol. 2008;74(24):7759-7766. DOI: 10.1128/AEM.01178-08

[25] Ma JK, Drake PM, Christou P. The production of recombinant pharmaceutical proteins in plants. Nat Rev Genet. 2003;4(10):794-805. DOI: 10.1038/nrg1177

[26] Cino J. High-yield protein production from Pichia pastoris yeast: a protocol for benchtop fermentation. Am Biotechnol Lab. 1999;17: 10-21.

[27] Martinez JL, Liu LF, Petranovic D, Nielsen J. Pharmaceutical protein production by yeast: towards production of human blood proteins by microbial fermentation. Curr Opin Biotechnol. 2012;23(6):965-971. DOI: 10.1016/j.copbio.2012.03.011

[28] Egel-Mitani M, Andersen AS, Diers II, Hach M, Thim L, Hastrup S, Vad K. Yield improvement of heterologous peptides expressed in yps1-disrupted *Saccharomyces cerevisiae* strains. Enzyme Microb Technol. 2000;26(9-10):671-677. DOI: 10.1016/s0141-0229(00)00158-7

[29] Song GY, Chung BH. Overproduction of human parathyroid hormone by fed-batch culture of a *Saccharomyces cerevisiae* mutant lacking yeast aspartic protease 3. Process Biochem. 1999;35(5):503-508. DOI: 10.1016/S0032-9592(99)00097-7

[30] Hadiji-AbbesN,MartinM,BenzinaW, Karray-Hakim H, Gergely C, Gargouri A, Mokhad-Gargouri R. Extraction and purification of hepatitis B virus-like M particles from a recombinant *Saccharomyces cerevisiae* strain using alumina powder. J Virol Methods.

2013;187:132-137. DOI: 10.1016/j.jviromet.2012.09.023

[31] Zhang L, Liu J, Lu J, Yan B, Song L, Li, L, Cui F, Zhang G, Wang F, Liang X, Xu A. Antibody response to revaccination among adult non-responders to primary Hepatitis B vaccination in China. Hum Vaccin Immunother. 2015a;11:2716-2722. DOI: 10.1080/21645515.2015.1045172

[32] King TH, Shanley CA, Guo Z, Bellgrau D, Rodell T, Furney S, Henao-Tamayo M, Orme IM. GI-19007, a novel *Saccharomyces cerevisiae*-based therapeutic vaccine against tuberculosis. Clin Vaccine Immunol. 2017;24:e00245–e00217. DOI: 10.1128/CVI.00245-17

[33] Kaslow DC, Shiloach J. Production, purification and immunogenicity of a malaria transmission-blocking vaccine candidate: TBV25H expressed in yeast and purified using nickel-NTA agarose. Biotechnology (N Y). 1994;12:494-499. DOI: 10.1038/nbt0594-494

[34] Fazlalipour M, Keyvani H, Monavari SH, Mollaie HR. Expression, purification and immunogenic description of a hepatitis C virus recombinant CoreE1E2 protein expressed by yeast Pichia pastoris. Jundishapur J Microbiol. 2015;8:e17157. DOI: 10.5812/jjm.8(4)2015.17157

[35] Jacobs E, Rutgers T, Voet P, Dewerchin M, Cabezon T, de Wilde M. Simultaneous synthesis and assembly of various hepatitis B surface proteins in *Saccharomyces cerevisiae*. Gene 1989;80:279-291. DOI: 10.1016/0378-1119(89)90292-8

[36] Kim HJ, Kim SY, Lim SJ, Kim JY, Lee SJ, Kim HJ. One-step chromatographic purification of human papillomavirus type 16 L1 protein from *Saccharomyces cerevisiae*. Protein Expression Purif. 2010;70:68-74. DOI: 10.1016/j.pep.2009.08.005

[37] Kim HJ, Lee JY, Kang HA,Lee Y, Park EJ, Kim HJ. Oral immunization with whole yeast producing viral capsid antigen provokes a stronger humoral immune response than purified viral capsid antigen. Lett Appl Microbiol. 2014;58:285-291. DOI: 10.1111/lam.12188

[38] Malak A, Baronian K, Kunze G. Blastobotrys (Arxula) adeninivorans: a promising alternative yeast for biotechnology and basic research. Yeast. 2016;33:535-547. DOI: 10.1002/yea.3180

[39] van Ooyen AJ, Dekker P, Huang M, Olsthoorn MMA, Jacobs DI, Colussi PA, Taron CH. Heterologous protein production in the yeast Kluyveromyces lactis. FEMS Yeast Res. 2006;6:381-392. DOI: 10.1111/j.1567-1364.2006.00049.x

[40] Vieira Gomes AM, Souza Carmo T, Silva Carvalho L, Mendonça Bahia F, Skorupa Parachin N. Comparison of yeasts as hosts for recombinant protein production. Microorganisms. 2018;6(2):38. OI: 10.3390/microorganisms6020038

[41] Goffeau A, Barrell BG, Bussey H, Davis RW, Dujon B, Feldmann H, Galibert F, Hoheisel JD, Jacq C, Johnston M, Louis EJ, Mewes HW, Murakami Y, Philippsen P, Tettelin H, Oliver SG. Life with 6000 genes. Science. 1996;274(5287):546, 563-547. DOI: 10.1126/science.274.5287.546

[42] Giga-Hama Y, Kumagai H. Expression system for forein genes using the fission yeast *Schizosaccharomyces pombe*. Biotechnology and Applied Biochemistry. 2000;30(3):235-244. DOI: 10.1111/j.1470-8744.1999.tb00776.x

[43] Drăgan CA, Buchheit D, Bischoff D, Ebner T, Bureik M. Glucuronide production by whole-cell biotransformation using genetically engineered fission yeast

*Schizosaccharomyces pombe*. Drug Metab Dispos. 2010;38(3):509-515. doi: 10.1124/dmd.109.030965

[44] Wood V, Gwilliam R, Rajandream MA, Lyne M, Lyne R, Stewart A, Sgouros J, Peat N, Hayles J, Baker S, Basham D, Bowman S, Brooks K, Brown D, Brown S, Chillingworth T, Churcher C, Collins M, Connor R, Cronin A, Davis P, Feltwell T, Fraser A, Gentles S, Goble A, Hamlin N, Harris D, Hidalgo J, Hodgson G, Holroyd S, Hornsby T, Howarth S, Huckle EJ, Hunt S, Jagels K, James K, Jones L, Jones M, Leather S, McDonald S, McLean J, Mooney P, Moule S, Mungall K, Murphy L, Niblett D, Odell C, Oliver K, O'Neil S, Pearson D, Quail MA, Rabbinowitsch E, Rutherford K, Rutter S, Saunders D, Seeger K, Sharp S, Skelton J, Simmonds M, Squares R, Squares S, Stevens K, Taylor K, Taylor RG, Tivey A, Walsh S, Warren T, Whitehead S, Woodward J, Volckaert G, Aert R, Robben J, Grymonprez B, Weltjens I, Vanstreels E, Rieger M, Schäfer M, Müller-Auer S, Gabel C, Fuchs M, Düsterhöft A, Fritzc C, Holzer E, Moestl D, Hilbert H, Borzym K, Langer I, Beck A, Lehrach H, Reinhardt R, Pohl TM, Eger P, Zimmermann W, Wedler H, Wambutt R, Purnelle B, Goffeau A, Cadieu E, Dréano S, Gloux S, Lelaure V, Mottier S, Galibert F, Aves SJ, Xiang Z, Hunt C, Moore K, Hurst SM, Lucas M, Rochet M, Gaillardin C, Tallada VA, Garzon A, Thode G, Daga RR, Cruzado L, Jimenez J, Sánchez M, del Rey F, Benito J, Domínguez A, Revuelta JL, Moreno S, Armstrong J, Forsburg SL, Cerutti L, Lowe T, McCombie WR, Paulsen I, Potashkin J, Shpakovski GV, Ussery D, Barrell BG, Nurse P. The genome sequence of *Schizosaccharomyces pombe*. Nature. 2002;415:871-880. DOI: 10.1038/nature724

[45] Hoffman CS, Wood V, Fantes PA. An ancient yeast for young geneticists: A primer on the *Schizosaccharomyces pombe* model system. Genetics. 2015;201:403-423. DOI: 10.1534/genetics.115.181503

[46] Darby RA, Cartwright SP, Dilworth MV, Bill RM. Which yeast species shall I choose? *Saccharomyces cerevisiae* versus Pichia pastoris (review). Methods Mol Biol. 2012;866:11-23. DOI: 10.1007/978-1-61779-770-5_2

[47] Shams H. Recent developments in veterinary vaccinology. Vet J. 2005;170(3):289-299. DOI: 10.1016/j.tvjl.2004.07.004

[48] Fang Y, Liu MQ, Chen L, Zhu ZG, Zhu RZ, Hu Q. Rabies post-exposure prophylaxis for a child with severe allergic reaction to rabies vaccine. Hum Vaccin Immunother 2016;12:1802-1804. DOI: 10.1080/21645515.2016.1143158

[49] Mager WH, Winderickx J. Yeast as a model for medical and medicinal research. Trends Pharmacol Sci. 2005;26:265-273. DOI: 10.1016/j.tips.2005.03.004

[50] Kumar R, Kumar P. Yeast-based vaccines: New perspective in vaccine development and application. FEMS Yeast Research. 2019;19. DOI: 10.1093/femsyr/foz007

[51] Rosenfeld AB, Racaniello VR. Hepatitis C virus internal ribosome entry site-dependent translation in Saccharomyces cerevisiae is independent of polypyrimidine tract-binding protein, poly(rC)-binding protein 2, and La protein 1. J Virol. 2005;79:10126-10137. DOI: 10.1128/JVI.79.16.10126-10137.2005

[52] Valenzuela P, Medina A, Rutter WJ, Ammerer G, Hall BD. Synthesis and assembly of hepatitis B virus surface antigen particles in yeast. Nature. 1982;298:347-350. DOI: 10.1038/298347a0

[53] Blanco R, Carrasco L, Ventoso I. Cell killing by HIV-1 protease. J Biol Chem.

2003;278:1086-1093. DOI: 10.1074/jbc.M205636200

[54] Kapoor P, Lavoie BD, Frappier L. EBP2 plays a key role in Epstein-Barr virus mitotic segregation and is regulated by aurora family kinases. Mol Cell Biol. 2005;25:4934-4945. DOI: 10.1128/MCB.25.12.4934-4945.2005

[55] Zhang X, Xin L, Li S, Fang M, Zhang J, Xia N, Zhao Q. Lessons learned from successful human vaccines: delineating key epitopes by dissecting the capsid proteins. Hum Vaccine Immunotherapeutics 2015;11:1277-1292. DOI: 10.1080/21645515.2015.1016675

[56] Nussenzweig V, Barr P, inventors; New York University, Chiron Corporation, assignee. Vaccine against the sporozoite stage of malaria. Patent US4997647A. 1991.

[57] Karumuthil-Melethil S, Gudi R, Johnson BM, Perez N, Vasu C. Fungal βglucan, a Dectin-1 ligand, promotes protection from type 1 diabetes by inducing regulatory innate immune response. J Immunol 2014;193:3308-3321. DOI: 10.4049/jimmunol.1400186

Chapter 5

# Checkpoint Control of DNA Repair in Yeast

*Michael Fasullo*

## Abstract

Budding yeast has been a model organism for understanding how DNA damage is repaired and how cells minimize genetic instability caused by arresting or delaying the cell cycle at well-defined checkpoints. However, many DNA damage insults are tolerated by mechanisms that can both be error-prone and error-free. The mechanisms that tolerate DNA damage and promote cell division are less well-understood. This review summarizes current information known about the checkpoint response to agents that elicit both the G2/M checkpoint and the intra-S phase checkpoint and how cells adapt to unrepaired DNA damage. Tolerance to particular bulky DNA adducts and radiomimetic agents are discussed, as well as possible mechanisms that may control phosphatases that deactivate phosphorylated proteins.

**Keywords:** DNA damage tolerance, checkpoint, budding yeast, phosphatase

## 1. Introduction

DNA repair involves the recognition and excision of DNA damage followed by template-directed DNA synthesis using an undamaged strand (for reviews see [1]). Major repair mechanisms include base excision repair (BER, [2]), nucleotide excision repair (NER, [3]), and double-strand break (DSB) repair [4]; In budding yeast, homologous recombination (HR) is the preferred pathway for repair of DSBs. When the DNA replication apparatus bypasses DNA lesions on either the leading or lagging strand, single-strand gaps are created, and the resulting repair of gapped sister chromatids is referred to as postreplication repair (PRR); postreplication repair pathway involves DNA synthesis by low fidelity polymerases or template switch mechanisms. These studies have determined the identity of multiple components of these DNA repair pathways and demonstrated a remarkable conservation between both "simple" eukaryotes, such as budding yeast, and higher eukaryotes, including humans. Yeast studies, therefore, have a direct impact on understanding the molecular basis of inheritable DNA repair deficiencies in humans, many of which are associated with cancer (for review, see [5]).

One unifying theme of DNA repair is the redundancy of DNA repair mechanisms for specific DNA lesions. In budding yeast, the pathway choice may depend on the context of the DNA lesion, the stage of the cell cycle, and the ploidy of the strain. For example, DSBs can be repaired by non-homologous end joining (NHEJ), single-strand annealing (SSA), and HR [4]). HR is enhanced [6, 7] while NHEJ is suppressed in MATa/MATα diploid strains [8] where the HR repair can use an undamaged homolog as a repair template. Secondly, specific DNA base lesions, such as abasic sites can be repaired by both bases excision repair (BER) or nucleotide excision repair (NER, [2, 9, 10]). Thirdly, postreplication repair (PPR) pathways can be both error-prone and error-free [11]. While some DNA lesions are repaired by redundant mechanisms, others, such as inter-strand DNA cross-links that impede DNA polymerases, require components of multiple pathways, including NER, postreplication repair, and HR [12].

Cells adapt to unrepaired DNA lesions and rely on DNA damage tolerance mechanisms to maintain viability. For example, exposure to 150 $J/m^2$ UV, generates $3 \times 10^4$ cyclobutane pyrimidine dimers (CPD) per yeast cell, or approximately one UV-induced dimer per 400 bp of yeast DNA [13]. The pyrimidine-pyrimidone (6–4) photoproduct is also abundant but eightfold less present after exposure to UVB (280–320) [14]. The efficiency of CPD repair depends on the surrounding chromatin and whether the damaged strand is transcribed (for review, [15]). Since not all UV-induced damage is repaired within the period of a single cell cycle, cell viability depends on DNA damage tolerance and adaptation.

The purpose of this review is to summarize mechanisms by which checkpoint activation and DNA damage tolerance confer resistance for particular DNA lesions, and to summarize more recent data concerning complex carcinogen-associated lesions. The importance of this topic is underscored by observations that tolerance of DNA damage may reduce the efficacy of chemotherapeutic drugs, such as cisplatin, while increasing genetic instability. We present studies that suggest that DNA damage tolerance can be influenced by multiple factors, including the nutritional status of the cell and signaling from both the Target of Rapamycin (TOR) and the protein kinase A (PKA) pathways.

The orchestration of DNA repair pathways is especially critical when the replication fork progression is blocked by bulky DNA adducts [16]. Stalled replication forks can generate DNA secondary structures that trigger genome instability. While particular mechanisms to bypass such adducts, such as template switching and translesion DNA synthesis, may be error free, there is a risk that toxic recombination intermediates can either impede DNA replication progression or lead to replication fork collapse [17]. Replication-associated DNA breaks, in turn, may re-initiate replication on an undamaged chromosome or chromatid, referred to as break-induced replication (BIR, [18]). Chromosome breaks, if unrepaired, can be aberrantly rejoined forming dicentric chromosomes, leading to further breakage and instability, often referred to as the breakage-fusion-bridge cycle, a phenomena suggested to account for gross genome rearrangements in yeast [19] and in cancer cells [20, 21].

To suppress genetic instability and facilitate DNA repair, cell cycle checkpoints trigger arrest at defined stages in the cell cycle to ensure that DNA damage is repaired before the damage is replicated or inherited in the next cell cycle (for review, see [22]). These checkpoints are referred to as the G1-S checkpoint, the intra S checkpoint, and the G2/M checkpoint. In brief, phosphoinositide three-kinase-related kinase (PI3K)-like kinases, referred to as apical or sensor kinases, initiate signaling after recruitment to DNA damage or stalled replication forks, checkpoint kinases, referred to as effector or downstream kinases, then amplify and transmit the checkpoint signal, and effectors that catalyze covalent protein modifications [23], resulting in activation or degradation of cellular targets. Besides ensuring that the cell cycle is delayed so that adequate time is available for DNA repair [24], effectors also modify DNA repair proteins [25, 26]), upregulate the synthesis of deoxynucleotides (dNTPs, [27]), regulate transport of tRNA from the nucleus to and from the cytoplasm, trigger autophagy [28], regulate histone levels [29], and cross-talk with other stress-induced pathways to ensure survival [30]. The totality of the response is generally referred to as the DNA damage response (DDR) [22]. Checkpoint recovery occurs once the DNA damage is repaired or the replication block has been circumvented [31]; however, adaptation to persistent DNA lesions, such as DSBs, can also occur. In either case, the phosphorylated and activated checkpoint proteins are either dephosphorylated or degraded and subsequently rendered inactive. Depending on the DNA damage and time period of exposure, peak activation occurs within 2–4 hours after acute DNA damage exposure [32], with simultaneous upregulation of dNTP levels and DNA damage-inducible genes. Repression of late origins of replication, inhibition of replication, and prevention of anaphase can last for additional hours [33], while adaptation can be observed after 12 hrs. The presence of DNA repair foci, such as Rad51, can thus last hours after the initiation of the DNA damage insult. Thus, aspects of the DNA damage response can persist hours after the initial genomic insult and after DNA repair is completed.

## 2. Checkpoint activation initiated by DSBs

Checkpoint activation triggered by DSBs, and subsequent recovery or adaptation has been extensively studied in strains containing either uncapped telomeres or chromosomal DSBs that cannot be repaired by HR. An unrepaired DSB occurs when HO endonuclease cleaves the recognition sequence at the MAT locus but silent mating type locus has been deleted [34]. Uncapped telomeres occur when either the Cdc13-Stn1-Ten1 (CST) complex or the Ku complex, composed of yKu70 and yKu80, is defective. At restrictive (elevated) temperatures in either *cdc13* or *yku70* mutants, extensive tracts of single-stranded DNA complex are generated [35]. While two DNA ends are revealed by a single unrepaired DSB at the MAT locus, at the non-permissive temperature in *cdc13* mutants single-stranded DNA is revealed at the telomeres of sixteen chromosomes, thus amplifying the DNA damage signal.

A single DSB occurring in G1 does not trigger cell cycle arrest at the G1/S or intra S phase checkpoint [36], but instead the cell cycle progresses through S phase and into G2 phase, where cells arrest. Repair of DSBs can occur at any time in the cell cycle by NHEJ; however, in budding yeast, NHEJ is favorable when the single-strand overhangs are short [37]. However, DSBs, will trigger a partial DNA damage response in G1 cells [38], and recombination proteins, such as Rad51 and Rad54, are still induced [39] and Rad55 is phosphorylated [38] .

The orchestration of checkpoint signaling has been well described in current reviews [40] and is briefly summarized (**Figure 1**). Mre11/Rad50/Xrs2 (MRX) and Tel1 (ataxia telangiectasia mutated (ATM) ortholog) bind to the ends of the DSB, which facilitates the juxtaposition of the ends of the breaks. NHEJ requires yKu70 and yKu80. However, if NHEJ is not successful, cyclin dependent kinase I (Cdk1 or Cdc28), which has high activity in G2, phosphorylates Sae2 and the 5′ to 3′ exonuclease Dna2 [41]. Sae2 phosphorylation activates the Mre11 endonuclease activity that ejects yKu70 from the ends of the DSB. Together with Sgs1/Dna2 and Exo1, the ends are further degraded in a 5′ to 3′ direction. NHEJ and resection require chromatin remodeling factors, including the Ino80 complex [42, 43], Rsc complex [44], and Fun30 [45, 46]. Resection is generally slow and proceeds at 1–2 nucleotide per minute [40]. Resection of the ends reveals single stranded DNA (ssDNA), which is then coated by single strand binding protein (RPA), which serves as a general sensor for DNA damage. The RPA-coated ssDNA is a binding site for Ddc2-Mec1 (ataxia telangiectasia mutated and rad3-related (ATR ortholog). Rad24/Rfc facilitates the binding of trimeric Rad17/Mec3/Ddc1 (9–1-1) protein which recognizes the junction between the single stranded DNA and the double-stranded DNA [47]. Thus, ssDNA serves as a general signal for checkpoint signaling [48].

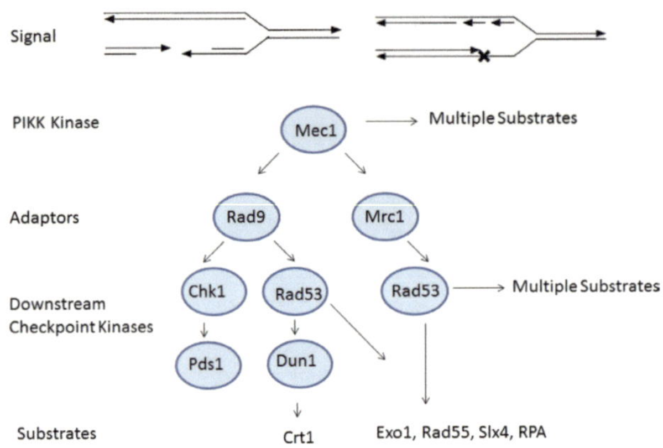


**Figure 1.**
*A pathway for checkpoint pathway commences with a DNA damage signal that triggers the PIKK kinase, Mec1. Downstream checkpoint kinases are activated, as facilitated by the adaptors Rad9 and Mrc1. Kinase substrates are identified for Rad53 and Dun1, but both Mec1 and Rad53 phosphorylate multiple substrates that are not shown.*

Mec1, a sensor or apical serine/threonine kinase phosphorylates downstream kinases, DNA repair proteins, and histones, preferably at SQ/TQ sites [49]. Both Mec1 and Tel1 phosphorylate histone γ-H2A for ~50 kb on either side of the DSB, which serves to recruit other checkpoint protein, such as the adaptor, Rad9 (53BP1 ortholog). Mec1 regulates checkpoint signaling by autophosphorylation on the S1964 residue [50] and phosphorylation of Ddc2, which destabilizes unbound Ddc2 and limits the amount of bound Ddc2-Mec1. Mec1 also phosphorylates Exo1 [51], which limits the amount of single-stranded DNA that could serve as a signal for checkpoint activation. Thus, Mec1's activity serves to not only activate downstream kinases but also dampen the checkpoint response.

Rad9, as an adaptor protein and 53BP1 ortholog, is required to bring the effector (transducer) kinases in contact with Mec1. Rad9 binding to chromatin is mediated by its BRCT and tudor domains that interact with phosphorylated and trimethylated histone H3, respectively [52]. While histone phosphorylation is induced by DNA damage, Dot1-mediated histone H3 methylation is constitutive [53]. Localization to damaged DNA is facilitated by binding to Rtt107/Dbp11. Both Mec1 and Cdk1 phosphorylate Rad9 on separate domains [40]. In turn, oligomers of phosphorylated Rad9 bind to Rad53 and facilitate Mec1-mediated phosphorylation. A Rad53 phosphorylated heterodimer then autophosphorylates; the hyper-phosphorylated Rad53 can, in turn, rapidly diffuse throughout the nucleus and phosphorylate multiple substrates, including Dun1 and Asf1. Similarly, Rad9 facilitates Mec1-mediated Chk1 phosphorylation; the activated Chk1 phosphorylates Pds1, which prevents its degradation by the anaphase promoting complex (APC). In turn, sister chromatid cohesion is maintained and anaphase is prevented [54]. Activated Rad53 also inhibits the APC from degrading securin [54]. A Rad53-mediated pathway inhibits Cdc5, a polo-like kinase that functions in mitotic exit through the regulation of spindle pole body separation [30]. While Pds1 phosphorylation can also be triggered by the Mad2-mediated spindle checkpoints, Rad53 phosphorylation is only triggered by DNA damage or stalled replication forks. While the single mutants are partially defective in DNA damage-induced G2 arrest, *rad53 pds1* and *rad53 dun1* double mutants are fully deficient [55].

Partial to full checkpoint activation will also occur when DNA damage processing is rendered less efficient. For example, mating-type switching in a *rad1* mutant, defective in removal of 3′ non-homologous ends, will trigger a checkpoint-dependent cell cycle delay [56]. Interestingly, Rad53 phosphorylation was not abundant in the *rad1* mutant; however, the cell cycle delay was not observed in the *rad9* strain, and was shortened in the *mad3* mutants, defective in spindle checkpoint. The authors speculated that checkpoint activation occurred when H2A phosphorylation extended through centromeric chromatin, triggering a spindle pole checkpoint response [56]. These studies indicate that spindle pole checkpoints also participate in the DNA damage response, depending on the context of the DSB.

While the G2 checkpoint is critical for HR repair of DSBs, Mec1 and Mec1-signaling pathway also phosphorylate additional DNA repair functions that facilitate DSB repair and damage incurred by radiomimetic agents [40]. For example, Mec1 phosphorylates Rad51 [57] and Rad55 [58]. Phosphorylation of Rad51 enhances its activity and is required for resistance to recombinagens, such as methyl methane sulfonate (MMS) [57]. In addition, Mec1 phosphorylates Slx4, which binds to Rad1/Rad10 and facilitates single-strand annealing by cleaving non-homologous tails [59]. The studies indicate that the checkpoint pathway directly phosphorylates repair proteins to enhance their function. While there are many proteins that are phosphorylated in response to DNA damage [60, 61], the functional significance of the phosphorylation of many of these proteins has yet to be determined (**Table 1**).

| Protein Phosphorylated | Kinase | Effect | Reference |
|---|---|---|---|
| Checkpoint Signaling | | | |
| Rad9 | Mec1 /Tel1 | Rad53 docking and Rad9 multimerization | [40, 60] |
| Rad53 | Mec1 | Activation of Rad53 autophosphorylation | [40, 60] |
| Rad53 | Tel1 | Activation of Rad53 | [40, 60] |
| Chk1 | Mec1 | Phosphorylation of Pds1 | [30, 54] |
| Pds1 | Chk1 | APC-associated degradation of Pds1 is inhibited | [54, 60] |
| Mec1-Ddc2 | Mec1 | Attenuation of Mec1 kinase activity | [50] |
| Dun1 | Rad53 | Activation of Dun1 kinase activity | [40] |
| Nucleases | | | |
| Sae2 | Cdk1 | Cell cycle regulation limiting resection to G2/M | [41] |
| Dna2 | Cdk1 | Cell cycle regulation limiting resection to G2/M | [62] |
| Exo1 | Mec1 | Inhibition of Exo1 5′-3′ exonuclease activity | [51] |
| Transcription inhibitors | | | |
| Crt1 | Dun1 | Crt1 phosphorylation leads to degradation, and subsequent Rnr transcriptional activation | [60, 63] |
| Protein Inhibitors | | | |
| Sml1 | Dun1 | Sml1 phosphorylation leads to degradation and release from Rnr1 subunit and subsequent increase in dNTPs | [27] |
| Dif1 | Dun1 | Allows for transport of RNR into the cytoplasm | [64] |
| Recombination Proteins | | | |
| Rad55 | Rad53 | Enhances recombination in rad5 mutants | [58] |
| Rad51 | Mec1/Rad53 | Enhances activity | [57] |
| Rev1 | Mec1 | Facilitates binding to ssDNA | [26] |

| Protein Phosphorylated | Kinase | Effect | Reference |
|---|---|---|---|
| Resolvases | | | |
| Yen1 | Cdk1 | Inhibits function in S phase by transportation to the cytoplasm | [62] |
| Mus81/Mms4 | Cdk1, Dbf4 Cdc7 | Regulation of cleavage of Holliday and branch junctions | [62, 65] |
| Helicases | | | |
| Pif1 | Rad53 | Inhibits fork unwinding, promotes DNA damage tolerance by HR | [66] |
| Rrm3 | Rad53 | Inhibits fork unwinding | [66] |
| Srs2 | Cdk1 | Promotes adaptation by removal of Rad51 filaments | [62, 67] |

**Table 1.**
*Proteins phosphorylated by DDR.*

## 3. Checkpoint recovery and adaptation from double-strand break

Once cells have repaired the DSB, recovery involves reversal of protein modifications and chromatin restoration. While the DNA damage may no longer be present, protein modifications are still present that signal checkpoint activation. To inactivate the G2/M checkpoint and resume division, Rad53 must be dephosphorylated. Two phosphatases involved in the inactivation of Rad53 include phosphorylated versions of the type 2C protein phosphatases (PP2C), Ptc2 and Ptc3 [68–70]; these phosphatases are also involved in inactivating other stress induced pathways, such as the Hog1-mediated osmotic stress induced pathway [71], while Ptc2 dephosphorylates Cdk1. Casein kinase II (Ck2) phosphorylates Ptc2, which specifically binds to the Rad53 FHA1 domains [72]. Interestingly, CK2 mutants are more defective in adaptation than *ptc2* mutants, suggesting that CK2 may control additional genes involved in adaptation [68].

Pph3, a member of the PP4 family, is important in maintaining full recovery; the triple mutant (*ptc2, ptc3, pph3*) is severely defective in DSB repair when the repair pathway is slow [70]. This may be partially explained by observations that Pph3 functions to dephosphorylate γ-H2A, which serves as a signal for activation of checkpoint proteins, cohesins, and chromatin remodelers [73]. However, the mechanism by with chromatin associated gamma γ-H2A is fully dephosphorylated is still being explored.

Chromatin restoration requires Asf1 and Caf1 which reassemble chromatin on DNA (Kim and Haber [74]). Asf1 binds histone H3 triggering acetylation by the histone acetyltransferase, Rtt109, and further ubiquitylation by Rtt101 [75]. This, in turn promotes the binding of the histone H3 and H4 heterodimer by Caf1. Interestingly, Asf1 also functions to bind Rad53, thus serving a role to sequester dephosphorylated Rad53. Thus Asf1 functions both in reassembling chromatin and stabilizing dephosphorylated Rad53 [75].

If a DSB is not repaired, cells will either resume the cell cycle or die. The resumption of the cell cycle is referred to as adaptation. Similar to recovery, adaptation involves both chromatin remodeling and phosphatases that deactivate the Rad53 kinase and Cdk1 kinase. This adaptation is blocked in *yku80* mutants [69], deficient in NHEJ, and *cdc5-ad*, which is defective in mitotic exit. *Yku80* mutants exhibit twice the rate of resection of the DSBs, resulting in more single-stranded DNA and thus the potential for more Rad53 checkpoint signaling. This interpretation is supported by observations that overexpression of Ptc2 is sufficient to suppress the adaptation defect of both *yku70* and *yku80* [69]. However, the role of resection in checkpoint adaptation is complicated by the identification of chromatin remodelers, such as Fun30 [76–78], which are required for adaptation but enhance resection. One possibility is that Fun30-associated resection in γ -H2A-modified chromatin antagonizes the checkpoint protein Rad9 from binding and signaling downstream checkpoint effectors [78].

Additional genes function to remove recombination proteins from chromatin. Removal of Rad51 filaments is facilitated by the chromatin modifier Tid1 (Rdh54) [79] and the Srs2 helicase, the former is phosphorylated by Mec1 and the latter is phosphorylated by Cdk1 [67]. Both *rdh54* and *srs2* mutants are defective in adaptation (reference [79, 80]). These studies present additional evidence *MEC1* functions both in the triggering of checkpoint arrest as well as recovery from checkpoint arrest.

## 4. Uncapped telomeres, checkpoint activation, and adaptation

While single-stranded DNA is present at telomeres, it is normally "capped" by a RPA-like structure, referred to as Cdc13-Stn1-Ten1 (CST), and by Ku (yKu70/yKu80) complex [35]. During replication, Cdc13 is phosphorylated by Cdk1 and recruits telomerase [81]. Telomere ends are susceptible to nucleases in yeast mutants defective in proteins that bind to chromosome ends, such as *yku70*, and that are defective in recruiting telomerase, such as the *cdc13–1* mutant at the restrictive temperature. Pif1 helicase inhibits telomerase and leads to slow resection at the telomere end [82, 83]. Resection is also slowed by binding of Rif1 and Rap1, which bind specifically to single stranded telomere sequences and inhibit the binding of the checkpoint activators, RPA and Rad24 [84, 85]. In the *cdc13–1* mutant, resection is extensive and largely performed by Exo1, leading to ssDNA bound to RPA, the 9–1-1 complex, and Rad9. Similarly, in *yku70* mutants, ssDNA is generated, but it takes several generations for ssDNA to accumulate [35]. The 9–1-1 complex is apparently not involved in eliciting a checkpoint response but Chk1 activation is required for Exo1-mediated resection [86]. In *yku70* mutants, resected telomeres elicit both a spindle and DNA damage checkpoint activation. However, unlike HO-induced DSBs, Mec1 binding does not lead to rapid resection but rather an inhibition of resection through subsequent binding of Rad9 and Rad53 [87]). Resection of the telomere, in turn, may facilitate recombination or break-induced replication (BIR, [88]) using an undamaged chromosomal end as a template for replication to the end of the chromosome. BIR is facilitated by activated Pif1 [89]. Thus, checkpoint activation at uncapped telomeres enables alternative mechanisms of telomere lengthening.

Adaptation to shortened telomeres was first noted by Sandell and Zakian [90] and require CKII and Cdc5 [91]. CKII directly phosphorylates Ptc2, which is required for tolerating shortened telomeres [92]. In addition, phosphorylated Cdc13 can be dephosphorylated by Pph3/Psy3, resulting in the segregation of uncapped chromosomal ends [35]. Over-expression of Cdc5 also decreases Rad53 phosphorylation [93]. Thus, as in HO-induced DSBs, there are multiple phosphatases and kinases that modulate adaptation.

## 5. Intra-S phase checkpoint and stabilization of the replication fork

The purpose of the intra-S phase checkpoint is to maintain replication fork integrity so that replication can be completed; collapsed replication forks are a major source of genetic instability [94]. Replication forks stall because of limiting amounts of dNTPs or when DNA damage, resulting from a bulky adduct or cross-links, block progression of a high fidelity polymerase (for review, see [95]). The precise number of stalled forks to trigger the intra S-phase checkpoint is unknown [96]. Uncoupling of the helicase and DNA polymerase activity generates single strand gaps on both the leading and lagging strands. Checkpoint responses serve to maintain the stability of the replication fork in part by blocking the formation of toxic secondary DNA structures and replication fork reversal, degrading excessive histones, and inhibiting the firing of late replication forks so that replication can be resumed if stalled replication forks become permanently arrested [97]. In S phase checkpoint mutants, such as *rad53*, regressed replication forks, referred to as "chicken feet" structures, can be visualized [98]. Severe deficiencies can lead to mitotic catastrophe and subsequent lethality.

The extensive tracts of single-stranded DNA generated at stalled forks signal a checkpoint response. Similar, to checkpoint signaling at DSBs, the 9–1-1 complex is loaded and facilitates binding of Ddc2-Mec1. Rad18, which monoubiquitinates trimeric PCNA at K164 position at stalled replication forks, also monoubiquitinates the 9–1-1 complex leading to enhanced recruitment of Ddc2-Mec1 [99]. Checkpoint activation at stalled forks can also be facilitated by Elg1, which removes PCNA from stalled forks [100]. The Mrc1 (claspin) functions as an adaptor, analogous to Rad9, in the phosphorylation of Rad53; however, unlike Rad9, Mrc1 is associated with the replication forks [101]. Full checkpoint activation requires BLM homolog Sgs1 [102]. Rad53 phosphorylation in turn serves to promote histone degradation, inhibit late origin firing, and increase the levels of dNTPs. The inhibition of late origin firing maintains RPA and allows replication restart from other replication origins.

Deoxynucleotide levels (dNTPs) increase 7–8 fold after DNA damage by upregulating the activity of ribonucleotide reductase (Rnr) activity [103]. Upregulation of Rnr activity is achieved at the transcriptional, translational, and the posttranslational levels. At the transcriptional level, phosphorylated Rad53 activates Dun1 kinase, which deactivates Crt1 transcriptional repressor [63]. At the translational level, TRM9, which functions to methylate the uridine wobble base of tRNA-Arg (UCU) and tRNA-Glu (UUC), facilitates the translation of the Rnr1 transcript [104]; however, how checkpoint signaling enhances *TRM9* function is unclear. At the post-translational level, Dun1 inactivates the Sml1 protein inhibitor by phosphorylation [27]. Besides increasing the transcription of Rnr subunits, the Rnr inhibitor Sml1 and its paralog Dif1 are degraded, subsequently the Rnr subunits are shuttled to the cytoplasm where they form an active enzyme complex [64]. In addition to increasing the overall level of dNTPs, the Rnr3 large subunit forms an alternative ribonucleotide reductase complex that has relaxed dATP negative feedback regulation [105]. This ensures that adequate levels of dNTPs are available during times of unscheduled DNA synthesis.

While high levels of dNTPs facilitate replication fork progression [106]; abnormally high or low levels of dNTPs can promote genetic instability. High levels of dNTPs reduce the fidelity of polymerase epsilon [107]. Low levels of dNTPs correlate with hyper-recombination, as has been observed in *dun1* null mutants and in mec1 hypomorphs; these phenotypes can be suppressed by higher basal levels of dNTPs conferred by a *SML1* deletion [108]. These studies suggest that there is a range of dNTP concentrations that correlate with replication fork stability; however, the mechanisms by which higher dNTP levels decrease replication fork collapse are unclear.

In contrast to DSBs, where HR processes are facilitated, there are redundant mechanisms to prevent recombination at stalled forks; these mechanisms include disassembly of Rad51 filaments, helicases that abort recombination intermediates,

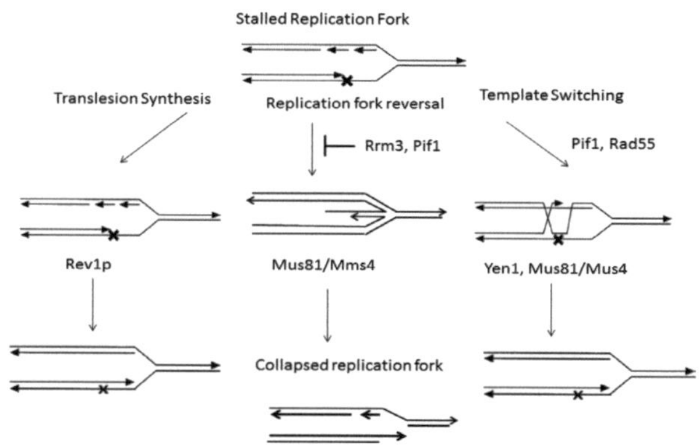


**Figure 2.**
*The pathways of tolerating DNA damage at a stalled fork are shown by the arrows. The 5′ to 3′ polarity of the DNA is designated by an arrow. The Rrm3 and Pif1 helicases inhibit replication fork reversal, while the Pif1 helicase promotes template switching. Phosphorylated Rev1 binds to single-stranded DNA and promotes replication bypass.*

and nucleases that degrade aberrant structures. PCNA is SUMOylated (SUMO-PCNA), facilitating the binding of the helicase Srs2, which suppresses recombination by disassembling Rad51 filaments (for review, see [109]. Recombination intermediates are aborted by the BLM ortholog Sgs1 helicase; indeed, the *sgs1 srs2* double mutants is not viable but viability is rescued when HR is defective [110]. Additional helicases, including Pif1 and Rrm3 can unwind and reverse "chicken feet" structures and reversed forks [66]. Nucleases, such as Exo1, function to degrade reversed forks, although excessive Exo1 activity can lead to replication fork collapse [62]. Finally, enzymes which cleave aberrant secondary structures, such as Yen1 and Mus4/Mus81, are inhibited or rendered less active by phosphorylation and Sumoylation (SUMO) [65]). Thus, there are multiple mechanisms that prevent aberrant structures from accumulating at replication blocks (**Figure 2**).

Generally, replication blocks that impede DNA polymerases can be bypassed by two mechanisms: 1) lesion bypass inserts a base opposite the replication block using error-prone or error-free translesion polymerases, and 2) template switch mechanisms [111] utilize recombination so that DNA polymerase bypasses DNA lesions on an undamaged template. Factors recruited to stalled forks would initially suggest that checkpoint signaling might favor lesion bypass by translesion polymerases. For example, Rad5 binding to stalled forks facilitates the recruitment of Rev1, even in the absence of DNA damage [112]. In addition, Rev1 is also phosphorylated by Mec1, which increases its affinity to ssDNA [26]. In vitro replication experiments have demonstrated that other error-prone polymerases can "jump start" replication, even without association of the replicative helicase [113]. However, there is no evidence that high levels of dNTPs would facilitate translesion synthesis mediated by error-prone polymerases, such as polζ [114], and the mechanism for jump start is unclear.

## 6. Bypass of single-strand gaps and replication blocks by template switch mechanisms

Template switch mechanisms also allow polymerases to bypass replication forks and resume DNA synthesis; these mechanisms are generally thought to occur on both leading and lagging strands. Template switching is orchestrated by proteins that modify the DNA polymerase processivity factor, PCNA. When the high fidelity polymerase stalls at the replication block, Rad18/Rad6 monoubiquitinates PCNA at the K164 position; monoubiquitinated PCNA can facilitate polymerase switching from to a translesion polymerase of lower fidelity and processivity. PCNA may further become polyubiquitinated at position K164 by combined action of Ubc13/Mms2/Rad5 (for review, see [115]). Rad5 also contains a helicase function that catalyzes replication fork reversal and is required for template switch mechanisms on the lagging strand [116]. While Rad5 does not require DNA damage at the stalled replication fork for recruitment [112], Rad5 over-expression can trigger genome instability [117]. The checkpoint signaling cascade, mediated by the Dun1 kinase, regulates Rad5 at the post-transcriptional level by destabilizing Rad5 mRNA [118]. These studies indicate that *RAD5* function is regulated.

However, checkpoint signaling may also facilitate template switch mechanisms. Rad53 is required for DNA damage-associated unequal SCE after exposure to MMS [119] and Rad53-mediated Rad55 phosphorylation confers enhanced MMS resistance when RAD5 is also defective [25]. The Rad9 checkpoint protein binds to persistent single strand gaps on the lagging strand, inhibiting the RecQ-like Sgs1 anti-recombination function. In addition Pif1, which is phosphorylated by Rad53, functions in template switching [120]. With longer term checkpoint-mediated G2 arrest, however, Rev1 protein levels accumulate [121, 122], suggesting that error-prone polymerases may serve as the ultimate backup in postreplication repair after error-free mechanisms have failed.

## 7. Choice of DNA damage tolerance pathway is influenced by the DNA lesion

Multiple tolerance pathways can confer resistance to particular types of DNA damage and the pathway preference depends on the DNA damaging agent. For example, MMS exposure generates by $^{7}$Me-Guanine and $^{3}$Me-Adenine lesions; while the $^{7}$Me-Guanine is mutagenic, the $^{3}$Me-Adenine blocks replication [123]. Replication bypass can occur by error-prone or error-free polymerases, or by template switching. While all three pathways are involved in bypass of $^{3}$Me-Adenine lesions [124], template switch mechanisms are preferred [125]. Checkpoint signaling facilitates template switch mechanisms after exposure to MMS [125, 126]. These studies suggest that template switch mechanisms may be the preferred pathway for bypassing particular lesions that block DNA replication.

The preference of template switch mechanisms or translesion pathways may depend on the efficiency of bypass and repair for large bulky adduct or cross-links. Particular UV-associated DNA cross-links are efficiently bypassed using either polε [127] or a two-step mechanism involving polε and polζ [128]. However, error-free bypass of 4–6 pyrmidine-pyrimidone lesions, present on a plasmid, occurs by template switch mechanisms after their introduction in a NER deficient yeast strain [129]. Likewise, 4-NQO induces bulky damage and stimulates template switch mechanisms [126]. These studies indicate that template switch mechanisms are likely used in error-free postreplication repair pathways [130].

## 8. Attenuation of the S phase checkpoint activation

In order for the cell cycle to resume and chromatids to separate the checkpoint activation needs to be downregulated and joint molecules need to be resolved. Once replication is completed, Mrc1 functions as an adaptor for Mec1-mediated checkpoint signaling is diminished since there are no more replication forks [33].

Resumption of the cell cycle is accomplished by dephosphorylating Rad53 [131]. However, single-strand gaps on sister chromatids can still function to trigger Rad9-mediated checkpoint signaling. To dampen Rad9's adaptor function in mediating Mec1 catalyzed Rad53 phosphorylation, competitive scaffolds compete with Rad9 binding to chromatin [132]. For example, the Mec1-mediated phosphorylation of Slx4 enables an association with Rtt107/Dpb11, which provides a competitive scaffold for the interaction of Rad9 with Dpb11 [133]. These mechanisms thus prevent Rad53 hyperphosphorylation.

Cleavage of joint DNA strands, or Holliday structures, is timed just before anaphase so that cleavage does not occur during S phase. Both kinases and phosphatases fine tune the timing of joint molecule cleavage. Cdk phosphorylates structure-specific nucleases Slx1/Slx4 and Mus81/Mus4 in late G2 and M phases respectively [134]. Whereas Mec1 phosphorylates and subsequently inactivates Yen1, Cdc14 dephosphorylates the inactivated form in mitosis, ensuring that joint molecules do not hinder sister chromatid division [134].

Similar to adaptions to DSBs, phosphatases deactivate Rad53 (**Figure 3**). These phosphatases include Pph3/Psy2 complex and Ptc1, 2. Interestingly, Pph3 directly interacts with Mec1/Ddc2 [135] at the replication fork, although the interaction does not rely on DNA damage [135]. Besides Rad53, other Mec1 substrates are likely dephosphorylated by Pph3, including phosphorylated Mec1. Thus Pph3 could potentially upregulate Mec1. However, the full range of Pph3 substrates is unknown [135].

Mutations in different phosphatases may confer sensitivities to different DNA damaging agents (**Table 2**). For example, *pph3* and *psy2* mutants are hypersensitive to phelomycin but not 4NQO, while *ptc2*, *ptc3* and *ptc2 ptc3* double mutants are not phleomycin sensitive and are not required for recovery from MMS-associated checkpoint delay [136, 142]. On the other hand *ptc2 ptc3* double mutants are hypersensitive to 4-NQO while *pph3* and *psy2* mutants are not sensitive. However, for particular agents, such as cisplatin, the triple *pph3 ptc2 ptc3* mutant, is synergistically more sensitive [140]. One idea is that phosphorylation of Rad53 is differentially patterned by particular DNA damaging agents, and that the phosphatases, Ptc2/Ptc3 and Pph3/Psy2 recognize different patterns [142]. This notion is supported by the identification of different MMS and 4-NQO associated Rad53 phosphorylation sites. The connection between Ptc2 and checkpoint activation is further strengthened by observations that over-expression of Ptc2 suppresses the lethality in a Rad53 dominant lethality mutant [69].

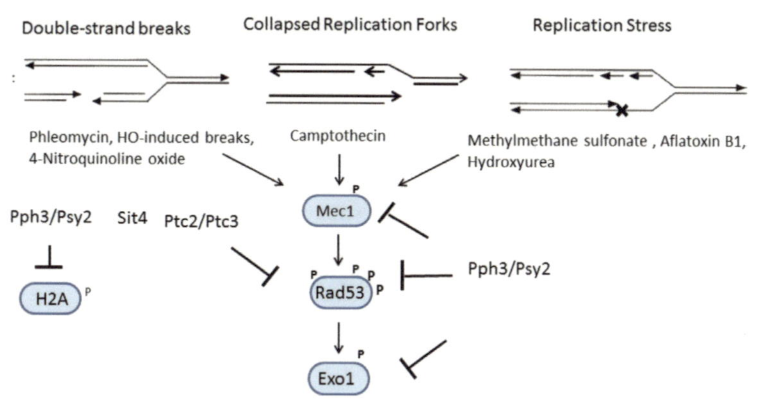


**Figure 3.**
*Double-strand breaks, collapsed replication forks, and replication stress lead to checkpoint activation. Activated Rad53 is dephosphorylated by a series of phosphatases, depending on the signal induced by the DNA damaging agent, shown below the DNA damage. The 5′ to 3′ polarity of the DNA is designated by an arrow. The substrates of the phosphatases Pph3/Psy2 and Ptc2/Ptc3 include Mec1, Rad53, and Exo1.*

| DNA damaging agent or environmental condition | DNA damage | Phosphatase Required for Resistance or Adaptation | Phenotype of Δ*ptc2 ptc3 pph3* | Reference |
|---|---|---|---|---|
| Restrictive temperature for *cdc13* mutant | Long tracts of ssDNA | Ptc2 | Unknown | [84] |
| HO Endonuclease | Double-strand break | Ptc2/Ptc3 | Completely deficient in adaptation | [70] |
| 4-Nitroquinoline oxide | Bulky adduct and oxidative damage | Ptc2/Ptc3, Sit4 | Unknown | [136, 137] |
| Phleomycin | Single-strand and double-strand breaks | Pph3, Ptc2/Ptc3 | Unknown | [136] |
| Methyl methanesulfonate | Major and minor groove alkylations | Pph3/Psy2 | Synergistically sensitive | [70] |
| Aflatoxin B1 | AFB1-N7-Gua, and AFB1 formamidopyrimidine | Pph3/Psy2 | Unknown | [138] |
| Ultraviolet radiation | Pyrimidine dimers and Pyrimidine-pyrimidone cross links | Not required | Moderate sensitivity | [70] |
| Hydroxyurea | Stalled DNA replication forks, double-strand breaks | Pph3/Psy2, Pph2 | Synergistically sensitive | [70, 139] |
| Camptothecin | Topo1 cross-link with DNA | Ptc2, Ptc3 | Synergistically sensitive | [70] |
| Cisplatin | DNA cross links | Pph3/Psy2 | Synergistically sensitive | [140] |
| Bleomycin | Single and double-strand break s | Pph3, Rts1 (regulator of Cdc55) | Unknown | [73, 141] |

**Table 2.**
*Phosphatases that function in checkpoint adaptation to specific DNA damaging agents.*

Tolerance to MMS-induced DNA damage includes reactivation of stalled replication forks, which depends on the level of Rad53 phosphorylation [143]. Pph3/Psy2 phosphatase is the principle phosphatase that deactivates Rad53. In the absence Pph3/Psy2 replication restart can occur; however late origins are used to complete DNA replication. Interestingly, downregulation of Rad53 phosphorylation by a HA-Rad53 or a *dot1* deletion confers higher levels of MMS resistance, although at the sake of more Rev1 foci and mutagenesis [144]. These studies would suggest that MMS-induced checkpoint activation is a double-edged sword; limiting MMS-induced mutation may come at the cost of toxic recombination intermediates.

While tolerance to MMS-induced DNA damage relies on dampening the checkpoint response, UV resistance heavily relies on checkpoint activation, as illustrated by observations that the *rad14 mec1* double mutant, defective in both NER and checkpoint signaling, is synergistically more UV sensitive [145]. In yeast, UV triggers the G1-S checkpoint when NER is functional, but unrepaired UV lesions trigger checkpoint responses in S and G2 cells [146]. Interestingly, chronic exposure to low dose UV does not elicit cell cycle arrest at the G2 checkpoint, suggesting that DNA replication machinery is not significantly impeded during chronic exposure [147].

P450-acitvated carcinogens may also elicit a strong DNA damage inducible effect. For example, aflatoxin B1 (AFB1), induces strong Rad53 activation in budding yeast, which generally occurs within two hours of exposure and then is gradually attenuated [148]). AFB1 exposure also upregulates the expression of DNA repair genes, including Rad51, Csm2, and Rad16 [149, 150]. Interestingly, AFB1 exposure elicits an S phase delay coinciding with the appearance of Rad51 foci [148]. This is consistent with AFB1 being a strong recombinagen but weak mutagen in yeast [151]. Interestingly, checkpoint signaling is required for stimulation of both AFB1-associated unequal sister chromatid recombination and mutation [152]. By profiling the yeast genome for AFB1 resistance using next generation sequencing, St. John *et al.* [138] identified both HR genes, including Rad54, Rad55, and Csm2, and those encoding error-prone polymerases. Similar to alkylated induced damage, the Csm2(Shu) complex favors an error-free template switch mechanism [153]; thus, csm2 mutants are deficient in sister chromatid recombination but exhibit higher frequencies of AFB1-associated mutations.

Genes the confer AFB1 resistance included *PSY3*, *CKB1* and *CKB2*, which function in DNA damage tolerance [138]. While the genes encoding the CKII substrates, Ptc2 and Ptc3, did not appear in the screen, the identification of both CKII and Pph3 suggest that tolerance to AFB1-associated DNA damage requires both phosphorylation and dephosphorylation of multiple proteins. The identity of these proteins may further elucidate how AFB1-associated DNA damage is tolerated.

Additional phosphatases that function in DNA damage tolerance include PP2A and PP2A-like phosphatases. These phosphatases are composed of catalytic subunits, such as Pph21 and Pph22, scaffolding subunits, and regulatory subunits, such as Cdc55 and Rts1. While a direct interaction with phosphorylated Rad53 has not been demonstrated, the PP2A phosphatase suppresses the checkpoint response after HU exposure [139]. While the identity of all of the PP2A substrates is unknown, PP2A is involved in both cytokinesis and mitosis [154]. Particular regulatory subunits are required for tolerance to different DNA damaging agents. For example, Rts1 is required for DNA damage tolerance after *rad51* cells are exposed to bleomycin [141] and Sit4, a PP2A-like phosphatase, is required for tolerance to 4NQO [137].

## 9. Nutrient sensing and the regulation of adaptation and the checkpoint response

One unifying theme in DNA damage tolerance to multiple types of DNA lesions is that nutrient sensing plays an important role in promoting downregulation of the checkpoint response. Deregulation of *IRA1* and *IRA2*, which control glucose-growth signaling, prevent adaptation to uncapped telomeres in *cdc13* strains [155]. Inhibition of TOR1 by rapamycin prevents adaptation and aneuploidy in rad52 diploid strains exposed to DNA damaging agents [156].

Nutrient sensing is also important in controlling the checkpoint response through type 2A protein phosphatases. In the presence of plentiful carbon and nitrogen, target of rapamycin (TORC1) activates Mec1-signaling pathway by inhibiting PP2A and PP2A-like phosphatases. PP2A activators include ceramide and S-adenosyl methionine (SAM) [139]. The effect of this signaling on the PP2C and PP4 phosphatases is unclear. Nonetheless, these studies illustrate that the DNA damage response requires an active growth signaling response [139]. Recent data also suggests that TORC1 inhibition results in lower levels of checkpoint proteins [157]. Thus, it may appear that TORC1 may be required for both checkpoint activation and for adaptation.

## 10. Concluding remarks

Adaptation to DNA damage is critical for cell survival. The simple, straightforward model is that DNA damage activates checkpoint signaling kinases and that phosphatases, which are constitutively expressed, serve to dephosphorylate and deactivate phosphorylated proteins. Once the DNA damage is repaired, the checkpoint signaling ceases and activated proteins are dephosphorylated. However, yeast can adapt to DNA damage caused by diverse damaging agents and individual phosphatases are controlled by different kinases. In addition, cells exposed to different DNA damaging agents exhibit distinct Rad53 phosphorylation patterns and rely more on particular phosphatases for checkpoint adaptation. The checkpoint pathway also autoregulates itself and dampens its signaling in coordination with Cdk1. Finally, adaptation to particular DNA damage requires TORC1 function, which senses nutrient abundance. Thus, simple models are likely complicated by the complexity of the checkpoint responses elicited by distinct DNA damaging agents.

## 11. Future directions and implications

Understanding DNA damage tolerance and repair will have a significant impact on elucidating the mechanisms by which DNA adducts cause mutations and genome instability. While DNA damage tolerance has been well-studied for particular types of DNA damage, such as that caused by UV and MMS, the mechanisms for DNA damage tolerance of more complex lesions is still in its infancy. The importance of studying DNA damage tolerance mechanisms for complex agents is underscored by the importance of cross-linking agents, such as cisplatin, in cancer chemotherapy. In addition, understanding how DNA damage is tolerated may have important impacts in assessing the efficacy of antifungal agents. Elucidating DNA damage tolerance mechanisms will also be important in understanding how mutations and genetic instabilities arise when cells are exposed to low doses of the DNA damaging agent. These studies should elucidate mechanisms on how cellular aging, ploidy, and cell type may affect DNA damage tolerance pathways.

## Author details

Michael Fasullo
State University of New York Polytechnic Institute, Albany, New York, United States

*Address all correspondence to: mfasullo@sunypoly.edu

## References

[1] Chatterjee N, Walker GC. Mechanisms of DNA damage, repair, and mutagenesis. Environ Mol Mutagen. 2017; 58(5):235-263. DOI: 10.1002/em.22087.

[2] Bauer NC, Corbett AH, Doetsch PW. The current state of eukaryotic DNA base damage and repair. Nucleic Acids Res. 2015; 43(21):10083-10101. DOI: 10.1093/nar/gkv1136.

[3] Schärer OD. Nucleotide excision repair in eukaryotes. Cold Spring Harb Perspect Biol. 2013; 5(10):a012609. DOI: 10.1101/cshperspect.a012609.

[4] Wright WD, Shah SS, Heyer WD. Homologous recombination and the repair of DNA double-strand breaks. J Biol Chem. 2018; 293(27):10524-10535. DOI: 10.1074/jbc.TM118.000372.

[5] Hoeijmakers JH. Genome maintenance mechanisms for preventing cancer. Nature. 2001;411(6835):366-374. DOI: 10.1038/35077232.

[6] Heude M, Fabre F. a/alpha-control of DNA repair in the yeast Saccharomyces cerevisiae: genetic and physiological aspects. Genetics. 1993; 133(3):489-498.

[7] Fasullo M, Dave P. Mating type regulates the radiation-associated stimulation of reciprocal translocation events in Saccharomyces cerevisiae. Mol Gen Genet. 1994; 243(1):63-70. DOI: 10.1007/BF00283877.

[8] Kegel A, Sjöstrand JO, Aström SU. Nej1p, a cell type-specific regulator of nonhomologous end joining in yeast. Curr Biol. 2001; 11(20):1611-1617. DOI: 10.1016/s0960-9822(01)00488-2.

[9] Torres-Ramos CA, Johnson RE, Prakash L, Prakash S. Evidence for the involvement of nucleotide excision repair in the removal of abasic sites in yeast. Mol Cell Biol. 2000;20(10):3522-3528. DOI: 10.1128/mcb.20.10.3522-3528.2000

[10] Kim N, Jinks-Robertson S. Abasic sites in the transcribed strand of yeast DNA are removed by transcription-coupled nucleotide excision repair. Mol Cell Biol. 2010; 30(13):3206-3215. DOI: 10.1128/MCB.00308-10.

[11] Gangavarapu V, Santa Maria SR, Prakash S, Prakash L. Requirement of replication checkpoint protein kinases Mec1/Rad53 for postreplication repair in yeast. mBio. 2011; 2(3):e00079–e00011. DOI: 10.1128/mBio.00079-11.

[12] Lehoczký P, McHugh PJ, Chovanec M. DNA interstrand cross-link repair in Saccharomyces cerevisiae. FEMS Microbiol Rev. 2007 31(2):109-133. DOI: 10.1111/j.1574-6976.2006.00046.x.

[13] Resnick MA, Setlow JK. Repair of pyrimidine dimer damage induced in yeast by ultraviolet light. J Bacteriol. 1972; 109(3):979-986. DOI: 10.1128/JB.109.3.979-986.1972.

[14] Hung KF, Sidorova JM, Nghiem P, Kawasumi M. The 6-4 photoproduct is the trigger of UV-induced replication blockage and ATR activation. Proc Natl Acad Sci U S A. 2020; 117(23):12806-12816. DOI: 10.1073/pnas.1917196117.

[15] Boiteux S, Jinks-Robertson S. DNA repair mechanisms and the bypass of DNA damage in Saccharomyces cerevisiae. Genetics. 2013;193(4):1025-1064. DOI: 10.1534/genetics.112.145219.

[16] Lehmann CP, Jiménez-Martín A, Branzei D, Tercero JA. Prevention of unwanted recombination at damaged replication forks. Curr Genet. 2020; 66(6):1045-1051. DOI: 10.1007/s00294-020-01095-7.

[17] Xu X, Lin A, Zhou C, Blackwell SR, Zhang Y, Wang Z, Feng Q, Guan R, Hanna MD, Chen Z, Xiao W. Involvement of budding yeast Rad5 in translesion DNA synthesis through physical interaction with Rev1. Nucleic Acids Res. 2016 Jun 20; 44(11):5231-5245. DOI: 10.1093/nar/gkw183.

[18] Kramara J, Osia B, Malkova A. Break-Induced Replication: The Where, The Why, and The How. Trends Genet. 2018; 34(7):518-531. DOI: 10.1016/j.tig.2018.04.002.

[19] Paek AL, Kaochar S, Jones H, Elezaby A, Shanks L, Weinert T. Fusion of nearby inverted repeats by a replication-based mechanism leads to formation of dicentric and acentric chromosomes that cause genome instability in budding yeast. Genes Dev. 2009; 23(24):2861-2875. DOI: 10.1101/gad.1862709.

[20] Kitada K, Yamasaki T. The complicated copy number alterations in chromosome 7 of a lung cancer cell line is explained by a model based on repeated breakage-fusion-bridge cycles. Cancer Genet Cytogenet. 2008; 185(1):11-19. DOI: 10.1016/j.cancergencyto.2008.04.005.

[21] Plug-DeMaggio AW, Sundsvold T, Wurscher MA, Koop JI, Klingelhutz AJ, McDougall JK. Telomere erosion and chromosomal instability in cells expressing the HPV oncogene 16E6. Oncogene. 2004; 23(20):3561-3571. DOI: 10.1038/sj.onc.1207388.

[22] Lanz MC, Dibitetto D, Smolka MB. DNA damage kinase signaling: checkpoint and repair at 30 years. EMBO J. 2019; 38(18):e101801. DOI: 10.15252/embj.2019101801.

[23] Zhou C, Elia AE, Naylor ML, Dephoure N, Ballif BA, Goel G, Xu Q, Ng A, Chou DM, Xavier RJ, Gygi SP, Elledge SJ. Profiling DNA damage-induced phosphorylation in budding yeast reveals diverse signaling networks. Proc Natl Acad Sci U S A. 2016; 113(26):E3667–E3675. DOI: 10.1073/pnas.1602827113.

[24] Weinert TA, Hartwell LH. The RAD9 gene controls the cell cycle response to DNA damage in Saccharomyces cerevisiae. Science. 1988; 241(4863):317-322. DOI: 10.1126/science.3291120.

[25] Herzberg K, Bashkirov VI, Rolfsmeier M, Haghnazari E, McDonald WH, Anderson S, Bashkirova EV, Yates JR 3rd, Heyer WD. Phosphorylation of Rad55 on serines 2, 8, and 14 is required for efficient homologous recombination in the recovery of stalled replication forks. Mol Cell Biol. 2006;26(22):8396-8409. DOI: 10.1128/MCB.01317-06.

[26] Sabbioneda S, Bortolomai I, Giannattasio M, Plevani P, Muzi-Falconi M. Yeast Rev1 is cell cycle regulated, phosphorylated in response to DNA damage and its binding to chromosomes is dependent upon MEC1. DNA Repair (Amst). 2007;6(1):121-127. DOI: 10.1016/j.dnarep.2006.09.002.

[27] Zhao X, Chabes A, Domkin V, Thelander L, Rothstein R. The ribonucleotide reductase inhibitor Sml1 is a new target of the Mec1/Rad53 kinase cascade during growth and in response to DNA damage. EMBO J. 2001;20(13):3544-3553. DOI: 10.1093/emboj/20.13.3544.

[28] Dotiwala F, Eapen VV, Harrison JC, Arbel-Eden A, Ranade V, Yoshida S, Haber JE. DNA damage checkpoint triggers autophagy to regulate the initiation of anaphase. Proc Natl Acad Sci U S A. 2013;110(1):E41–E49. DOI: 10.1073/pnas.1218065109.

[29] Singh RK, Kabbaj MH, Paik J, Gunjan A. Histone levels are regulated by phosphorylation and ubiquitylation-dependent proteolysis. Nat Cell Biol.

2009;11(8):925-933. DOI: 10.1038/ncb1903.

[30] Sanchez Y, Bachant J, Wang H, Hu F, Liu D, Tetzlaff M, Elledge SJ. Control of the DNA damage checkpoint by chk1 and rad53 protein kinases through distinct mechanisms. Science. 1999;286(5442):1166-1171. DOI: 10.1126/science.286.5442.1166.

[31] Clémenson C, Marsolier-Kergoat MC. DNA damage checkpoint inactivation: adaptation and recovery. DNA Repair (Amst). 2009 Sep 2;8(9):1101-1109. DOI: 10.1016/j.dnarep.2009.04.008.

[32] Yeung M, Durocher D. Srs2 enables checkpoint recovery by promoting disassembly of DNA damage foci from chromatin. DNA Repair (Amst). 2011;10(12):1213-1222. DOI: 10.1016/j.dnarep.2011.09.005.

[33] Moriel-Carretero M, Pasero P, Pardo B. DDR Inc., one business, two associates. Curr Genet. 2019; 65(2):445-451. DOI: 10.1007/s00294-018-0908-7.

[34] Lee SE, Pellicioli A, Malkova A, Foiani M, Haber JE. The Saccharomyces recombination protein Tid1p is required for adaptation from G2/M arrest induced by a double-strand break. Curr Biol. 2001;11(13):1053-1057. DOI: 10.1016/s0960-9822(01)00296-2.

[35] Dewar JM, Lydall D. Similarities and differences between "uncapped" telomeres and DNA double-strand breaks. Chromosoma. 2012;121(2):117-130. DOI: 10.1007/s00412-011-0357-2.

[36] Doksani Y, Bermejo R, Fiorani S, Haber JE, Foiani M. Replicon dynamics, dormant origin firing, and terminal fork integrity after double-strand break formation. Cell. 2009; 137(2):247-258. DOI: 10.1016/j.cell.2009.02.016.

[37] Daley JM, Wilson TE. Rejoining of DNA double-strand breaks as a function of overhang length. Mol Cell Biol. 2005 Feb;25(3):896-906. doi: 10.1128/MCB.25.3.896-906.2005. PMID: 15657419; PMCID: PMC544009.

[38] Janke R, Herzberg K, Rolfsmeier M, Mar J, Bashkirov VI, Haghnazari E, Cantin G, Yates JR 3rd, Heyer WD. A truncated DNA-damage-signaling response is activated after DSB formation in the G1 phase of Saccharomyces cerevisiae. Nucleic Acids Res. 2010;38(7):2302-2313. DOI: 10.1093/nar/gkp1222.

[39] Basile G, Aker M, Mortimer RK. Nucleotide sequence and transcriptional regulation of the yeast recombinational repair gene RAD51. Mol Cell Biol. 1992; 12(7):3235-3246. DOI: 10.1128/mcb.12.7.3235.

[40] Waterman DP, Haber JE, Smolka MB. Checkpoint Responses to DNA Double-Strand Breaks. Annu Rev Biochem. 2020; 89:103-133. DOI: 10.1146/annurev-biochem-011520-104722.

[41] Ira G, Pellicioli A, Balijja A, Wang X, Fiorani S, Carotenuto W, Liberi G, Bressan D, Wan L, Hollingsworth NM, Haber JE, Foiani M. DNA end resection, homologous recombination and DNA damage checkpoint activation require CDK1. Nature. 2004;431(7011):1011-1017. DOI: 10.1038/nature02964.

[42] van Attikum H, Fritsch O, Hohn B, Gasser SM. Recruitment of the INO80 complex by H2A phosphorylation links ATP-dependent chromatin remodeling with DNA double-strand break repair. Cell. 2004; 119(6):777-788. DOI: 10.1016/j.cell.2004.11.033.

[43] van Attikum H, Fritsch O, Gasser SM. Distinct roles for SWR1 and INO80 chromatin remodeling complexes at chromosomal double-strand breaks. EMBO J. 2007; 26(18):4113-4125. DOI: 10.1038/sj.emboj.7601835.

[44] Chambers AL, Downs JA. The RSC and INO80 chromatin-remodeling complexes in DNA double-strand break repair. Prog Mol Biol Transl Sci. 2012; 110:229-261. DOI: 10.1016/B978-0-12-387665-2.00009-2.

[45] Costelloe T, Louge R, Tomimatsu N, Mukherjee B, Martini E, Khadaroo B, Dubois K, Wiegant WW, Thierry A, Burma S, van Attikum H, Llorente B. The yeast Fun30 and human SMARCAD1 chromatin remodellers promote DNA end resection. 2012; Nature 489:581-584. DOI: org/10.1038/nature11353.

[46] Eapen VV, Sugawara N, Tsabar M, Wu WH, Haber JE. The *Saccharomyces cerevisiae* chromatin remodeler Fun30 regulates DNA end resection and checkpoint deactivation. Mol Cell Biol. 2012; 32(22):4727-4740. DOI: 10.1128/MCB.00566-12.

[47] Melo JA, Cohen J, Toczyski DP. Two checkpoint complexes are independently recruited to sites of DNA damage in vivo. Genes Dev. 2001; 15(21):2809-2821. DOI: 10.1101/gad.903501.

[48] Garvik B, Carson M, Hartwell L. Single-stranded DNA arising at telomeres in cdc13 mutants may constitute a specific signal for the RAD9 checkpoint. Mol Cell Biol. 1995; 15(11):6128-6138. DOI: 10.1128/mcb.15.11.6128.

[49] Traven A, Heierhorst J. SQ/TQ cluster domains: concentrated ATM/ATR kinase phosphorylation site regions in DNA-damage-response proteins. Bioessays 2005; 27:397-407. DOI://dx.doi.org/10.1002/bies.20204

[50] Memisoglu G, Lanz MC, Eapen VV, Jordan JM, Lee K, Smolka MB, Haber JE. $Mec1^{ATR}$ Autophosphorylation and $Ddc2^{ATRIP}$ Phosphorylation Regulates DNA Damage Checkpoint Signaling. Cell Rep. 2019; 28(4):1090-1102.e3. DOI: 10.1016/j.celrep.2019.06.068

[51] Morin I, Ngo HP, Greenall A, Zubko MK, Morrice N, Lydall D. Checkpoint-dependent phosphorylation of Exo1 modulates the DNA damage response. EMBO J. 2008; 27(18):2400-2410. DOI: 10.1038/emboj.2008.171.

[52] Blankley RT, Lydall D. A domain of Rad9 specifically required for activation of Chk1 in budding yeast. J Cell Sci. 2004; 117(4):601-608. DOI: 10.1242/jcs.00907.

[53] Conde F, Ontoso D, Acosta I, Gallego-Sánchez A, Bueno A, San-Segundo PA. Regulation of tolerance to DNA alkylating damage by Dot1 and Rad53 in Saccharomyces cerevisiae. DNA Repair (Amst). 2010; 9(10):1038-1049. DOI: 10.1016/j.dnarep.2010.07.003.

[54] Agarwal R, Tang Z, Yu H, Cohen-Fix O. Two distinct pathways for inhibiting pds1 ubiquitination in response to DNA damage. J Biol Chem. 2003; 278(45):45027-45033. DOI: 10.1074/jbc.M306783200.

[55] Gardner R, Putnam CW, Weinert T. RAD53, DUN1 and PDS1 define two parallel G2/M checkpoint pathways in budding yeast. EMBO J. 1999;18(11):3173-3185. DOI: 10.1093/emboj/18.11.3173.

[56] Lyndaker AM, Goldfarb T, Alani E. Mutants defective in Rad1-Rad10-Slx4 exhibit a unique pattern of viability during mating-type switching in Saccharomyces cerevisiae. Genetics. 2008; 179(4):1807-1821. DOI: 10.1534/genetics.108.090654.

[57] Flott S, Kwon Y, Pigli YZ, Rice PA, Sung P, Jackson SP. Regulation of Rad51 function by phosphorylation. EMBO Rep. 2011; 12(8):833-839. DOI: 10.1038/embor.2011.127.

[58] Bashkirov VI, King JS, Bashkirova EV, Schmuckli-Maurer J, Heyer WD. DNA repair protein Rad55

is a terminal substrate of the DNA damage checkpoints. Mol Cell Biol. 2000; 20(12):4393-4404. DOI: 10.1128/mcb.20.12.4393-4404.2000.

[59] Toh GW, Sugawara N, Dong J, Toth R, Lee SE, Haber JE, Rouse J. Mec1/Tel1-dependent phosphorylation of Slx4 stimulates Rad1-Rad10-dependent cleavage of non-homologous DNA tails. DNA Repair (Amst). 2010; 9(6):718-726. DOI: 10.1016/j.dnarep.2010.02.013.

[60] Zhou C, Elia AE, Naylor ML, Dephoure N, Ballif BA, Goel G, Xu Q, Ng A, Chou DM, Xavier RJ, Gygi SP, Elledge SJ. Profiling DNA damage-induced phosphorylation in budding yeast reveals diverse signaling networks. Proc Natl Acad Sci U S A. 2016 Jun 28;113(26):E3667-75. doi: 10.1073/pnas.1602827113. Epub 2016 Jun 13. PMID: 27298372; PMCID: PMC4932963.

[61] Breitkreutz A, Choi H, Sharom JR, Boucher L, Neduva V, Larsen B, Lin ZY, Breitkreutz BJ, Stark C, Liu G, Ahn J, Dewar-Darch D, Reguly T, Tang X, Almeida R, Qin ZS, Pawson T, Gingras AC, Nesvizhskii AI, Tyers M. A global protein kinase and phosphatase interaction network in yeast. Science. 2010 May 21;328(5981):1043-1046. DOI: 10.1126/science.1176495.

[62] Ubersax JA, Woodbury EL, Quang PN, Paraz M, Blethrow JD, Shah K, Shokat KM, Morgan DO. Targets of the cyclin-dependent kinase Cdk1. Nature 2003; 425:859-864.

[63] Huang M, Zhou Z, Elledge SJ. The DNA replication and damage checkpoint pathways induce transcription by inhibition of the Crt1 repressor. Cell. 1998; 94(5):595-605. DOI: 10.1016/s0092-8674(00)81601-3

[64] Lee YD, Wang J, Stubbe J, Elledge SJ. Dif1 is a DNA-damage-regulated facilitator of nuclear import for ribonucleotide reductase. Mol Cell. 2008; 32(1):70-80. DOI: 10.1016/j.molcel.2008.08.018.

[65] Ramos F, Villoria MT, Alonso-Rodríguez E, Clemente-Blanco A. Role of protein phosphatases PP1, PP2A, PP4 and Cdc14 in the DNA damage response. Cell Stress. 2019; 3(3):70-85. DOI:10.15698/cst2019.03.178

[66] Rossi SE, Ajazi A, Carotenuto W, Foiani M, Giannattasio M. Rad53-Mediated Regulation of Rrm3 and Pif1 DNA Helicases Contributes to Prevention of Aberrant Fork Transitions under Replication Stress. Cell Rep. 2015; 13(1):80-92. DOI: 10.1016/j.celrep.2015.08.073.

[67] Chiolo I, Carotenuto W, Maffioletti G, Petrini JH, Foiani M, Liberi G. Srs2 and Sgs1 DNA helicases associate with Mre11 in different subcomplexes following checkpoint activation and CDK1-mediated Srs2 phosphorylation. Mol Cell Biol. 2005;25(13):5738-5751. DOI:

[68] Marsolier MC, Roussel P, Leroy C, Mann C. Involvement of the PP2C-like phosphatase Ptc2p in the DNA checkpoint pathways of Saccharomyces cerevisiae. Genetics. 2000;154(4):1523-1532.

[69] Leroy C, Lee SE, Vaze MB, Ochsenbein F, Guerois R, Haber JE, Marsolier-Kergoat MC. PP2C phosphatases Ptc2 and Ptc3 are required for DNA checkpoint inactivation after a double-strand break. Mol Cell. 2003; 11(3):827-835. doi: 10.1016/s1097-2765(03)00058-3.

[70] Kim JA, Hicks WM, Li J, Tay SY, Haber JE. Protein phosphatases pph3, ptc2, and ptc3 play redundant roles in DNA double-strand break repair by homologous recombination. Mol Cell Biol. 2011; 31(3):507-516. DOI: 10.1128/MCB.01168-10.

[71] Johnson AJ, Veljanoski F, O'Doherty PJ, Zaman MS,

Petersingham G, Bailey TD, Münch G, Kersaitis C, Wu MJ. Molecular insight into arsenic toxicity via the genome-wide deletion mutant screening of Saccharomyces cerevisiae. Metallomics. 2016; 8(2):228-235. DOI: 10.1039/c5mt00261c.

[72] Pike BL, Tenis N, Heierhorst J. Rad53 kinase activation-independent replication checkpoint function of the N-terminal forkhead-associated (FHA1) domain. J Biol Chem. 2004;279(38):39636-33944. DOI: 10.1074/jbc.M405080200.

[73] Keogh MC, Kim JA, Downey M, Fillingham J, Chowdhury D, Harrison JC, Onishi M, Datta N, Galicia S, Emili A, Lieberman J, Shen X, Buratowski S, Haber JE, Durocher D, Greenblatt JF, Krogan NJ. A phosphatase complex that dephosphorylates gammaH2AX regulates DNA damage checkpoint recovery. Nature. 2006;439(7075):497-501. DOI: 10.1038/nature04384

[74] Kim JA, Haber JE. Chromatin assembly factors Asf1 and CAF-1 have overlapping roles in deactivating the DNA damage checkpoint when DNA repair is complete. Proc Natl Acad Sci U S A. 2009; 106(4):1151-1161. DI: 10.1073/pnas.0812578106.

[75] Tsabar M, Waterman DP, Aguilar F, Katsnelson L, Eapen VV, Memisoglu G, Haber JE. Asf1 facilitates dephosphorylation of Rad53 after DNA double-strand break repair. Genes Dev. 2016;30(10):1211-1224. DOI: 10.1101/gad.280685.116.

[76] Papamichos-Chronakis M, Krebs JE, Peterson CL. Interplay between Ino80 and Swr1 chromatin remodeling enzymes regulates cell cycle checkpoint adaptation in response to DNA damage. Genes Dev. 2006; 20(17):2437-2449. DOI: 10.1101/gad.1440206.

[77] Tong ZB, Ai HS, Li JB. The Mechanism of Chromatin Remodeler SMARCAD1/Fun30 in Response to DNA Damage. Front Cell Dev Biol. 2020 Sep 25;8:560098. DOI: 10.3389/fcell.2020.560098.

[78] Costelloe T, Louge R, Tomimatsu N, Mukherjee B, Martini E, Khadaroo B, Dubois K, Wiegant WW, Thierry A, Burma S, van Attikum H, Llorente B. The yeast Fun30 and human SMARCAD1 chromatin remodellers promote DNA end resection. Nature. 2012; 489(7417):581-584. DOI: 10.1038/nature11353.

[79] Ferrari M, Nachimuthu BT, Donnianni RA, Klein H, Pellicioli A. Tid1/Rdh54 translocase is phosphorylated through a Mec1- and Rad53-dependent manner in the presence of DSB lesions in budding yeast. DNA Repair (Amst). 2013; 12(5):347-355. DOI: 10.1016/j.dnarep.2013.02.004.

[80] Vaze MB, Pellicioli A, Lee SE, Ira G, Liberi G, Arbel-Eden A, Foiani M, Haber JE. Recovery from checkpoint-mediated arrest after repair of a double-strand break requires Srs2 helicase. Mol Cell. 2002; 10(2):373-385. DOI: 10.1016/s1097-2765(02)00593-2.

[81] Li S, Makovets S, Matsuguchi T, Blethrow JD, Shokat KM, Blackburn EH. Cdk1-dependent phosphorylation of Cdc13 coordinates telomere elongation during cell-cycle progression. Cell. 2009 Jan 9;136(1):50-61. doi: 10.1016/j.cell.2008.11.027. PMID: 19135888; PMCID: PMC2642970.

[82] Phillips JA, Chan A, Paeschke K, Zakian VA. The pif1 helicase, a negative regulator of telomerase, acts preferentially at long telomeres. PLoS Genet. 2015;11(4):e1005186. DOI: 10.1371/journal.pgen.1005186.

[83] Dewar JM, Lydall D. Pif1- and Exo1-dependent nucleases coordinate checkpoint activation following telomere uncapping. EMBO J.

2010 Dec 1;29(23):4020-34. doi: 10.1038/emboj.2010.267. Epub 2010 Nov 2. PMID: 21045806; PMCID: PMC3020640.

[84] Fontana GA, Reinert JK, Thomä NH, Rass U. Shepherding DNA ends: Rif1 protects telomeres and chromosome breaks. Microb Cell. 2018; 5(7):327-343. DOI: 10.15698/mic2018.07.639.37.

[85] Mattarocci S, Reinert JK, Bunker RD, Fontana GA, Shi T, Klein D, Cavadini S, Faty M, Shyian M, Hafner L, Shore D, Thomä NH, Rass U. Rif1 maintains telomeres and mediates DNA repair by encasing DNA ends. Nat Struct Mol Biol. 2017; 24(7):588-595. DOI: 10.1038/nsmb.3420.

[86] Maringele L, Lydall D. EXO1-dependent single-stranded DNA at telomeres activates subsets of DNA damage and spindle checkpoint pathways in budding yeast yku70Delta mutants. Genes Dev. 2002 Aug 1;16(15):1919-1933.

[87] Jia X, Weinert T, Lydall D. Mec1 and Rad53 inhibit formation of single-stranded DNA at telomeres of Saccharomyces cerevisiae cdc13-1 mutants. Genetics. 2004; 166(2):753-764. DOI: 10.1534/genetics.166.2.753.

[88] Holstein EM, Ngo G, Lawless C, Banks P, Greetham M, Wilkinson D, Lydall D. Systematic Analysis of the DNA Damage Response Network in Telomere Defective Budding Yeast. G3 (Bethesda). 2017; 7(7):2375-2389. DOI: 10.1534/g3.117.042283.

[89] Vasianovich Y, Harrington LA, Makovets S. Break-induced replication requires DNA damage-induced phosphorylation of Pif1 and leads to telomere lengthening. PLoS Genet. 2014; 10(10):e1004679. DOI: 10.1371/journal.pgen.1004679.

[90] Sandell LL, Zakian VA. Loss of a yeast telomere: arrest, recovery, and chromosome loss. Cell. 1993; 75(4):729-739. DOI: 10.1016/0092-8674(93)90493-a.

[91] Toczyski DP, Galgoczy DJ, Hartwell LH. CDC5 and CKII control adaptation to the yeast DNA damage checkpoint. Cell. 1997; 90(6):1097-1106. DOI: 10.1016/s0092-8674(00)80375-x.

[92] Mersaoui SY, Gravel S, Karpov V, Wellinger RJ. DNA damage checkpoint adaptation genes are required for division of cells harbouring eroded telomeres. Microb Cell. 2015; 2(10):394-405. DOI: 10.15698/mic2015.10.229.

[93] Vidanes GM, Sweeney FD, Galicia S, Cheung S, Doyle JP, Durocher D, Toczyski DP. CDC5 inhibits the hyperphosphorylation of the checkpoint kinase Rad53, leading to checkpoint adaptation. PLoS Biol. 2010 Jan 26;8(1):e1000286. DOI: 10.1371/journal.pbio.1000286.

[94] Myung K, Chen C, Kolodner RD. Multiple pathways cooperate in the suppression of genome instability in Saccharomyces cerevisiae. Nature. 2001; 411(6841):1073-1076. DOI 10.1038/35082608. PMID: 11429610.

[95] Marians KJ. Lesion Bypass and the Reactivation of Stalled Replication Forks. Annu Rev Biochem. 2018; 87:217-238. DOI: 10.1146/annurev-biochem-062917-011921.

[96] Shimada K, Pasero P, Gasser SM. ORC and the intra-S-phase checkpoint: a threshold regulates Rad53p activation in S phase. Genes Dev. 2002; 16(24):3236-3252. DOI: 10.1101/gad.239802.

[97] Tercero JA, Diffley JF. Regulation of DNA replication fork progression through damaged DNA by the Mec1/Rad53 checkpoint. Nature. 2001 Aug 2;412(6846):553-557. DOI: 10.1038/35087607.

[98] Lopes M, Foiani M, Sogo JM. Multiple mechanisms control chromosome integrity after replication fork uncoupling and restart at irreparable UV lesions. Mol Cell. 2006; 21(1):15-27. DOI: 10.1016/j.molcel.2005.11.015.0.

[99] Fu Y, Zhu Y, Zhang K, Yeung M, Durocher D, Xiao W. Rad6-Rad18 mediates a eukaryotic SOS response by ubiquitinating the 9-1-1 checkpoint clamp. Cell. 2008;133(4):601-611. DOI: 10.1016/j.cell.2008.02.050.

[100] Sau S, Liefshitz B, Kupiec M. The Yeast PCNA Unloader Elg1 RFC-Like Complex Plays a Role in Eliciting the DNA Damage Checkpoint. mBio. 2019; 10(3):e01159–e01119. DOI: 10.1128/mBio.01159-19.

[101] Osborn AJ, Elledge SJ. Mrc1 is a replication fork component whose phosphorylation in response to DNA replication stress activates Rad53. Genes Dev. 2003; 17(14):1755-1767. DOI: 10.1101/gad.1098303

[102] Bjergbaek L, Cobb JA, Tsai-Pflugfelder M, Gasser SM. Mechanistically distinct roles for Sgs1p in checkpoint activation and replication fork maintenance. EMBO J. 2005; 24(2):405-417. DOI: 10.1038/sj.emboj.7600511.

[103] Chabes A, Georgieva B, Domkin V, Zhao X, Rothstein R, Thelander L. Survival of DNA damage in yeast directly depends on increased dNTP levels allowed by relaxed feedback inhibition of ribonucleotide reductase. Cell. 2003; 112(3):391-401. DOI: 10.1016/s0092-8674(03)00075-8.

[104] Begley U, Dyavaiah M, Patil A, Rooney JP, DiRenzo D, Young CM, Conklin DS, Zitomer RS, Begley TJ. Trm9-catalyzed tRNA modifications link translation to the DNA damage response. Mol Cell. 2007; 28(5):860-870. DOI: 10.1016/j.molcel.2007.09.021

[105] Domkin V, Thelander L, Chabes A. Yeast DNA damage-inducible Rnr3 has a very low catalytic activity strongly stimulated after the formation of a cross-talking Rnr1/Rnr3 complex. J Biol Chem. 2002; 277(21):18574-18578. DOI: 10.1074/jbc.M201553200.

[106] Poli J, Tsaponina O, Crabbé L, Keszthelyi A, Pantesco V, Chabes A, Lengronne A, Pasero P. dNTP pools determine fork progression and origin usage under replication stress. EMBO J. 2012;31(4):883-894. doi: 10.1038/emboj.2011.470.

[107] Williams LN, Marjavaara L, Knowels GM, Schultz EM, Fox EJ, Chabes A, Herr AJ. dNTP pool levels modulate mutator phenotypes of error-prone DNA polymerase ε variants. Proc Natl Acad Sci U S A. 2015; 112(19):E2457–E2466. DOI: 10.1073/pnas.1422948112.

[108] Fasullo M, Tsaponina O, Sun M, Chabes A. Elevated dNTP levels suppress hyper-recombination in Saccharomyces cerevisiae S-phase checkpoint mutants. Nucleic Acids Res. 2010; 38(4):1195-1203. DOI: 10.1093/nar/gkp1064

[109] Arbel M, Liefshitz B, Kupiec M. How yeast cells deal with stalled replication forks. Curr Genet. 2020; 66(5):911-915. DOI: 10.1007/s00294-020-01082-y..

[110] Gangloff S, Soustelle C, Fabre F. Homologous recombination is responsible for cell death in the absence of the Sgs1 and Srs2 helicases. Nat Genet. 2000; 25(2):192-194. DOI: 10.1038/76055.

[111] Rupp WD, Howard-Flanders P. Discontinuities in the DNA synthesized in an excision-defective strain of Escherichia coli following ultraviolet irradiation. J Mol Biol. 1968 Jan 28;31(2):291-304. doi: 10.1016/0022-2836(68)90445-2

[112] Gallo D, Kim T, Szakal B, Saayman X, Narula A, Park Y, Branzei D, Zhang Z, Brown GW. Rad5 Recruits Error-Prone DNA Polymerases for Mutagenic Repair of ssDNA Gaps on Undamaged Templates. Mol Cell. 2019; 73(5):900-914.e9. DOI: 10.1016/j.molcel.2019.01.001.

[113] Guilliam TA, Yeeles JTP. Reconstitution of translesion synthesis reveals a mechanism of eukaryotic DNA replication restart. Nat Struct Mol Biol. 2020; 27(5):450-460. DOI: 10.1038/s41594-020-0418-4.

[114] Kochenova OV, Bezalel-Buch R, Tran P, Makarova AV, Chabes A, Burgers PM, Shcherbakova PV. Yeast DNA polymerase ζ maintains consistent activity and mutagenicity across a wide range of physiological dNTP concentrations. Nucleic Acids Res. 2017; 45(3):1200-1218. DOI: 10.1093/nar/gkw1149.

[115] Andersen PL, Xu F, Xiao W. Eukaryotic DNA damage tolerance and translesion synthesis through covalent modifications of PCNA. Cell Res. 2008; 18(1):162-173. DOI: 10.1038/cr.2007.114.

[116] Blastyák A, Pintér L, Unk I, Prakash L, Prakash S, Haracska L. Yeast Rad5 protein required for postreplication repair has a DNA helicase activity specific for replication fork regression. Mol Cell. 2007; 28(1):167-175. DOI: 10.1016/j.molcel.2007.07.030.

[117] Bryant EE, Šunjevarić I, Berchowitz L, Rothstein R, Reid RJD. Rad5 dysregulation drives hyperactive recombination at replication forks resulting in cisplatin sensitivity and genome instability. Nucleic Acids Res. 2019; 47(17):9144-9159. DOI: 10.1093/nar/gkz631.

[118] Hammet A, Pike BL, Heierhorst J. Posttranscriptional regulation of the RAD5 DNA repair gene by the Dun1 kinase and the Pan2-Pan3 poly(A)-nuclease complex contributes to survival of replication blocks. J Biol Chem. 2002; 277(25):22469-22474. DOI: 10.1074/jbc.M202473200.

[119] Fasullo M, Dong Z, Sun M, Zeng L. Saccharomyces cerevisiae RAD53 (CHK2) but not CHK1 is required for double-strand break-initiated SCE and DNA damage-associated SCE after exposure to X rays and chemical agents. DNA Repair (Amst). 2005; 4(11):1240-1251. DOI: 10.1016/j.dnarep.2005.06.006.

[120] Waters LS, Walker GC. The critical mutagenic translesion DNA polymerase Rev1 is highly expressed during G(2)/M phase rather than S phase. Proc Natl Acad Sci U S A. 2006; 103(24):8971-8976. DOI: 10.1073/pnas.0510167103.

[121] García-Rodríguez N, Wong RP, Ulrich HD. The helicase Pif1 functions in the template switching pathway of DNA damage bypass. Nucleic Acids Res. 2018; 46(16):8347-8356. DOI: 10.1093/nar/gky648.

[122] Beranek DT. Distribution of methyl and ethyl adducts following alkylation with monofunctional alkylating agents. Mutat Res. 1990; 231(1):11-30. DOI: 10.1016/0027-5107(90)90173-2.

[123] Plosky BS, Frank EG, Berry DA, Vennall GP, McDonald JP, Woodgate R. Eukaryotic Y-family polymerases bypass a 3-methyl-2′-deoxyadenosine analog in vitro and methyl methanesulfonate-induced DNA damage in vivo. Nucleic Acids Res. 2008 Apr;36(7):2152-2162. doi: 10.1093/nar/gkn058.

[124] Johnson RE, Yu SL, Prakash S, Prakash L. A role for yeast and human translesion synthesis DNA polymerases in promoting replication through 3-methyl adenine. Mol Cell Biol. 2007; 27(20):7198-7205. DOI: 10.1128/MCB.01079-07.

[125] Vanoli F, Fumasoni M, Szakal B, Maloisel L, Branzei D. Replication and recombination factors contributing to recombination-dependent bypass of DNA lesions by template switch. PLoS Genet. 2010; 6(11):e1001205. DOI: 10.1371/journal.pgen.1001205.

[126] Minca EC, Kowalski D. Multiple Rad5 activities mediate sister chromatid recombination to bypass DNA damage at stalled replication forks. Mol Cell. 2010; 38(5):649-661. DOI: 10.1016/j.molcel.2010.03.020.

[127] Yu SL, Johnson RE, Prakash S, Prakash L. Requirement of DNA polymerase eta for error-free bypass of UV-induced CC and TC photoproducts. Mol Cell Biol. 2001; 21(1):185-1888. DOI: 10.1128/MCB.21.1.185-188.2001

[128] Johnson RE, Haracska L, Prakash S, Prakash L. Role of DNA polymerase eta in the bypass of a (6-4) TT photoproduct. Mol Cell Biol. 2001; 21(10):3558-3563. DOI: 10.1128/MCB.21.10.3558-3563.2001.

[129] Zhang H, Lawrence CW. The error-free component of the RAD6/RAD18 DNA damage tolerance pathway of budding yeast employs sister-strand recombination. Proc Natl Acad Sci U S A. 2005 Nov 1;102(44):15954-15959. DOI: 10.1073/pnas.0504586102.

[130] Xu X, Blackwell S, Lin A, Li F, Qin Z, Xiao W. Error-free DNA-damage tolerance in Saccharomyces cerevisiae. Mutat Res Rev Mutat Res. 2015; 764:43-50. DOI: 10.1016/j.mrrev.2015.02.001

[131] Jablonowski CM, Cussiol JR, Oberly S, Yimit A, Balint A, Kim T, Zhang Z, Brown GW, Smolka MB. Termination of Replication Stress Signaling via Concerted Action of the Slx4 Scaffold and the PP4 Phosphatase. Genetics. 2015; 201(3):937-949. DOI: 10.1534/genetics.115.181479.

[132] Dibitetto D, Ferrari M, Rawal CC, Balint A, Kim T, Zhang Z, Smolka MB, Brown GW, Marini F, Pellicioli A. Slx4 and Rtt107 control checkpoint signalling and DNA resection at double-strand breaks. Nucleic Acids Res. 2016; 44(2):669-682. DOI: 10.1093/nar/gkv1080.

[133] Ohouo PY, Bastos de Oliveira FM, Liu Y, Ma CJ, Smolka MB. DNA-repair scaffolds dampen checkpoint signalling by counteracting the adaptor Rad9. Nature. 2013;493(7430):120-124. DOI: 10.1038/nature11658

[134] Campos A, Clemente-Blanco A. Cell Cycle and DNA Repair Regulation in the Damage Response: Protein Phosphatases Take Over the Reins. Int J Mol Sci. 2020; 21(2):446. DOI: 10.3390/ijms21020446.

[135] Hustedt N, Seeber A, Sack R, Tsai-Pflugfelder M, Bhullar B, Vlaming H, van Leeuwen F, Guénolé A, van Attikum H, Srivas R, Ideker T, Shimada K, Gasser SM. Yeast PP4 interacts with ATR homolog Ddc2-Mec1 and regulates checkpoint signaling. Mol Cell. 2015; 57(2):273-289. DOI: 10.1016/j.molcel.2014.11.016.

[136] Heideker J, Lis ET, Romesberg FE. Phosphatases, DNA damage checkpoints and checkpoint deactivation. Cell Cycle. 2007; 6(24):3058-3064. DOI: 10.4161/cc.6.24.5100.

[137] Douville J, David J, Fortier PK, Ramotar D. The yeast phosphotyrosyl phosphatase activator protein, yPtpa1/Rrd1, interacts with Sit4 phosphatase to mediate resistance to 4-nitroquinoline-1-oxide and UVA. Curr Genet. 2004 Aug;46(2):72-81. doi: 10.1007/s00294-004-0513.

[138] St John N, Freedland J, Baldino H, Doyle F, Cera C, Begley T, Fasullo M. Genome Profiling for Aflatoxin $B_1$ Resistance in *Saccharomyces cerevisiae* Reveals a Role for the CSM2/SHU

Complex in Tolerance of Aflatoxin $B_1$-Associated DNA Damage. G3 (Bethesda). 2020; 10(11):3929-3947. DOI: 10.1534/g3.120.401723.

[139] Ferrari E, Bruhn C, Peretti M, Cassani C, Carotenuto WV, Elgendy M, Shubassi G, Lucca C, Bermejo R, Varasi M, Minucci S, Longhese MP, Foiani M. PP2A Controls Genome Integrity by Integrating Nutrient-Sensing and Metabolic Pathways with the DNA Damage Response. Mol Cell. 2017; 67(2):266-281.e4. DOI: 10.1016/j.molcel.2017.05.027.

[140] Jain D, Patel N, Shelton M, Basu A, Roque R, Siede W. Enhancement of cisplatin sensitivity by NSC109268 in budding yeast and human cancer cells is associated with inhibition of S-phase progression. Cancer Chemother Pharmacol. 2010; 66(5):945-952. DOI: 10.1007/s00280-010-1246-1248.

[141] Aouida M, Eshrif A, Ramotar D. Yeast Lacking the PP2A Phosphatase Regulatory Subunit Rts1 Sensitizes *rad51* Mutants to Specific DNA Damaging Agents. Front Genet. 2019; 10:1117. DOI: 10.3389/fgene.2019.01117.

[142] Travesa A, Duch A, Quintana DG. Distinct phosphatases mediate the deactivation of the DNA damage checkpoint kinase Rad53. J Biol Chem. 2008; 283(25):17123-17130. DOI: 10.1074/jbc.M801402200.

[143] O'Neill BM, Szyjka SJ, Lis ET, Bailey AO, Yates JR 3rd, Aparicio OM, Romesberg FE. Pph3-Psy2 is a phosphatase complex required for Rad53 dephosphorylation and replication fork restart during recovery from DNA damage. Proc Natl Acad Sci U S A. 2007; 104(22):9290-9295. DOI: 10.1073/pnas.0703252104.

[144] Conde F, Ontoso D, Acosta I, Gallego-Sánchez A, Bueno A, San-Segundo PA. Regulation of tolerance to DNA alkylating damage by Dot1 and Rad53 in Saccharomyces cerevisiae. DNA Repair (Amst). 2010; 9(10):1038-1049. DOI: 10.1016/j.dnarep.2010.07.003

[145] Gangavarapu V, Santa Maria SR, Prakash S, Prakash L. Requirement of replication checkpoint protein kinases Mec1/Rad53 for postreplication repair in yeast. mBio. 2011; 2(3):e00079–e00011. DOI: 10.1128/mBio.00079-11.

[146] Neecke H, Lucchini G, Longhese MP. Cell cycle progression in the presence of irreparable DNA damage is controlled by a Mec1- and Rad53-dependent checkpoint in budding yeast. EMBO J. 1999; 18(16):4485-4497. DOI: 10.1093/emboj/18.16.4485.

[147] Hishida T, Kubota Y, Carr AM, Iwasaki H. RAD6-RAD18-RAD5-pathway-dependent tolerance to chronic low-dose ultraviolet light. Nature. 2009; 457(7229):612-615. DOI: 10.1038/nature07580.

[148] Fasullo M, Chen Y, Bortcosh W, Sun M, Egner PA. Aflatoxin B(1)-Associated DNA Adducts Stall S Phase and Stimulate Rad51 foci in Saccharomyces cerevisiae. J Nucleic Acids. 2010 Dec 2;2010:456487. DOI: 10.4061/2010/456487.

[149] Guo Y, Breeden LL, Fan W, Zhao LP, Eaton DL, Zarbl H. Analysis of cellular responses to aflatoxin B(1) in yeast expressing human cytochrome P450 1A2 using cDNA microarrays. Mutat Res. 2006; 593(1-2):121-142. DOI: 10.1016/j.mrfmmm.2005.07.001.

[150] Keller-Seitz MU, Certa U, Sengstag C, Würgler FE, Sun M, Fasullo M. Transcriptional response of yeast to aflatoxin B1: recombinational repair involving RAD51 and RAD1. Mol Biol Cell. 2004; 15(9):4321-4336. DOI: 10.1091/mbc.e04-05-0375.

[151] Sengstag C, Weibel B, Fasullo M. Genotoxicity of aflatoxin B1: evidence

for a recombination-mediated mechanism in Saccharomyces cerevisiae. Cancer Res. 1996; 56:5457-5465

[152] Fasullo M, Sun M, Egner P. Stimulation of sister chromatid exchanges and mutation by aflatoxin B1-DNA adducts in Saccharomyces cerevisiae requires MEC1 (ATR), RAD53, and DUN1. Mol Carcinog. 2008 Aug;47(8):608-15. DOI: 10.1002/mc.20417.

[153] Godin S, Wier A, Kabbinavar F, Bratton-Palmer DS, Ghodke H, Van Houten B, VanDemark AP, Bernstein KA. The Shu complex interacts with Rad51 through the Rad51 paralogues Rad55-Rad57 to mediate error-free recombination. Nucleic Acids Res. 2013; 41(8):4525-4534. DOI: 10.1093/nar/gkt138.

[154] Moyano-Rodriguez Y, Queralt E. PP2A Functions during Mitosis and Cytokinesis in Yeasts. Int J Mol Sci. 2019 Dec 30;21(1):264. DOI: 10.3390/ijms21010264.

[155] Wood MD, Sanchez Y. Deregulated Ras signaling compromises DNA damage checkpoint recovery in S. cerevisiae. Cell Cycle. 2010; 9(16):3353-3363. DOI: 10.4161/cc.9.16.12713.

[156] Vydzhak O, Bender K, Klermund J, Busch A, Reimann S, Luke B. Checkpoint adaptation in recombination-deficient cells drives aneuploidy and resistance to genotoxic agents. DNA Repair (Amst). 2020; 95:102939. DOI: 10.1016/j.dnarep.2020.102939.

[157] Miyamoto I, Ozaki R, Yamaguchi K, Yamamoto K, Kaneko A, Ushimaru T. TORC1 regulates the DNA damage checkpoint via checkpoint protein levels. Biochem Biophys Res Commun. 2019; 510(4):629-635. DOI: 10.1016/j.bbrc.2019.02.010.

Chapter 6

# The Budding Yeast *Saccharomyces cerevisiae* as a Valuable Model Organism for Investigating Anti-Aging Compounds

*Yanni Sudiyani, Muhammad Eka Prastya, Roni Maryana, Eka Triwahyuni and Muryanto*

## Abstract

*Saccharomyces cerevisiae*, the budding yeast was long history as industrial baker's yeast due to its ability to produce numerous product such as ethanol, acetate, industrial bakers etc. Interestingly, this yeast was also important tools for studying biological mechanism in eukaryotic cells including aging, autophagy, mitochondrial response etc. *S. cerevisiae* has arisen as a powerful chemical and genetic screening platform, due to a rapid workflow with experimental amenability and the availability of a wide range of genetic mutant libraries. Calorie restriction (CR) as the reduction of nutrients intake could promote yeast longevity through some pathways such as inhibition of nutrient sensing target of rapamycin (TOR), serine–threonine kinase (SCH9), protein adenylate cyclase (AC), protein kinase A (PKA) and ras, reduced ethanol, acetic acid and apoptotic process. In addition, CR also induces the expression of antioxidative proteins, sirtuin2 (Sir2), autophagy and induction of mitochondrial yeast adaptive response. Three methods, spotting test; chronological life span (CLS) and replicative life span (RLS) assays, have been developed to study aging in *S. cerevisiae*. Here, we present strategies for pharmacological anti-aging screens in yeast, discuss common pitfalls and summarize studies that have used yeast for drug discovery.

**Keywords:** *Saccharomyces cerevisiae*, anti-aging, calorie restriction, spotting test, chronological life span, replicative life span

## 1. Introduction

The budding yeast *Saccharomyces cerevisiae* is unicellular eukariotic fungi that divide asexually by budding. This particular yeast cells has an individual cell size of 5–10 μm and are pigmented, which cream color emerged in surface-grown colonies (**Figure 1**). Taxonomically, *S. cerevisiae* belongs to division of Ascomycota, class of Saccharomycetes, family of Saccharomycetaceae and genus of Saccharomyces. Yeast *S. cerevisiae* breaks down glucose in the medium through aerobic respiration in presence of oxygen. If oxygen is absent, the yeast will then undergo through anaerobic fermentation. This yeasts is widely distributed in the natural environment,

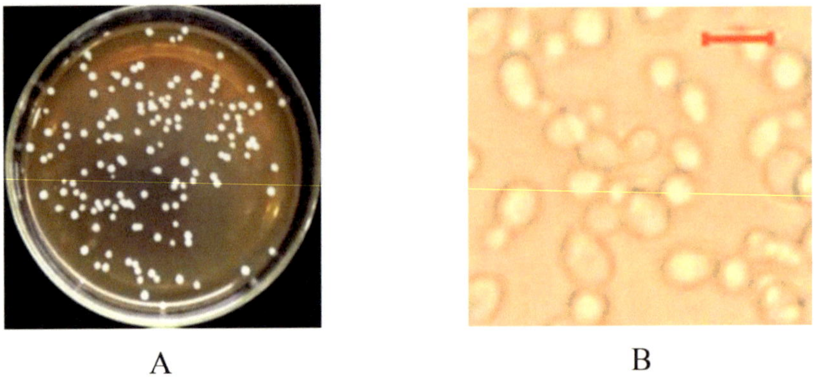


**Figure 1.**
*The budding yeast* S. cerevisiae, *(A) cells colony in agar plate medium; (B) cells morphology observed under microscope observation with 1000x magnification. Bars represent 5 μm. (picture permitted from [1], originally published in IOP Conf. Series: Earth and Environmental Science. doi: 10.1088/1755-1315/299/1/012059).*

including soil, water, plants, animals, and insects. Notably, *S. cerevisiae* are eukaryotic cells that contain all major organelles that are also common to animal cells like nucleus, mitochondria, endoplasmic reticulum, vacuole, golgi apparatus, cytoskeleton and many others organelles [2].

Scientist stated that *S. cerevisiae* was the first eukaryotic genome that was completely sequenced in 1996. In fact, the chromosomes of *S. cerevisiae* is compactly organized on 16 chromosomes and the genome has about 12.156.677 base pairs with 6.275 genes along with about 5.800 are believed to be functionally genes. Interestingly, it is estimated that this particular yeast have at least 31% of its genes homologous with that of humans belonging [3, 4]. As for, mating between haploid cells must occur to return to the diploid state. Both of haploid and diploid phases are morphologically similar, but with larger cells for diploid. In the asexual reproduction, bud grows to reach the size of the mother cell while nuclear division happens [5]. The separation would process after a nucleus is passed to the daughter cells with the smaller size particularly than mother cells. *S. cerevisiae* has a long history of uses in the area of food processing as leavening for bread and as a fermenter of wine production and alcoholic beverages, along with as a vitamin supplement due to it contains 50% protein and is a rich source of niacin, folic acid and B vitamins. In addition, *S. cerevisiae* is also clearly stated as the most ideal eukaryotic cells for biological studies. The incredible power of yeast genetics has made a legendary tools and is the envy of those who work with higher eukaryotic organisms [4, 6].

On the other hand, aging is a multifaceted process of accumulation of cellular, molecular and organ damage, leading to loss of function and increased vulnerability to disease following with the death. Indeed, there is a profound overlap between cellular and molecular pathways that influence aging and those linked to neurodegeneration, cancer, metabolic syndrome, and cardiovascular disorders. Therefore, recent efforts have emerged at the identification of compounds that decelerate the aging process and thus may act as a preventive measure that collectively ameliorates age-related diseases [7]. In fact, studies of aging in mammalian cells are limited by the long lifespan of common model organisms. Rats and mice live 3–5 years and primates up to 40. Nevertheless, aging studies, particularly in rodents, have been highly informative, of the prospective understanding of the genetic factors for modulating longevity [6]. Alternatively, a second approach that has dramatically accelerated aging research is the use of invertebrate model organisms, which age more rapidly and are readily amenable to environmental and genetic manipulation. Even though a variety of organisms have been investigated, a majority of studies have employed worms (*Caenorhabditis elegans*) [8], fruit flies (*Drosophila melanogaster*) [9], or yeast (*S. cerevisiae*) [10]. The use of model organisms at least in part due to its conserve some pathways that exist in all of these creatures, besides similarly in cellular or molecular events. One of the most popular conserve pathway is the nutrient-sensing pathway namely target of rapamycin (TOR) that exist in all of model organism starting from yeast to mammals [11].

The budding yeast *S. cerevisiae* has served as a model of organism and cellular aging for more than 50 years. In this particular yeast, many pathways that are relevant for aging and disease in humans are well conserved, including nutrient signaling, DNA repair mechanisms, cell cycle regulation, protein folding, lipostasis, mitochondrial homeostasis, stress response, secretion, proteostasis, and regulated cell death [6, 7]. In addition, the molecular mechanisms for cellular aging are also conserved from budding yeast to humans, primarily in particular condition with the nutrient depletion in growth medium, which retards aging and prolongs the lifespan. The conserved protein kinase, TORC1 (target of rapamycin complex 1), that is activated by nutrients and insulin, is a key regulator of the lifespan from eukaryotes organisms. The TORC1 inhibitor, namely rapamycin, retards aging and extends the lifespan of various species from yeast to mammals. In addition, this yeast has the advantages of speediness, simplicity, low cost, and good reproducibility. Furthermore, the genome DNA is not complex and has high homology with mammals. Indeed, all of DNA sequences are known and molecular mechanism research handling is relatively easy to perform, as well [6, 12, 13]. Some recent studies reported the utilization of *S.* cerevisase as a promising model organism to unravel anti-aging mechanism of compounds derived from various sources. Carmona-Gutierrez et al. [14] has shown anti-aging activity of flavonoid 4,4-dimethoxychalcone in *S. cerevisiae* by promoting authopagy mechanism. Lin et al., [15] reported anti-aging activity of Cucurbitacin B through regulating autophagy and oxidative stress in *S. cerevisiae*. In addition, Sudharsan et al., [16] also succeded to uncover the anti-aging mechanism of Astaxanthin in this yeast by decreasing oxidative stress and apoptosis mechanisms.

In this brief chapter, we discussed the utilization of *S. cerevisiae* as a model organism in drug discovery research, and how this simple eukaryote has been employed not only as a production vehicle due to its capability to produce some products i.e. wine, alcoholic beverage, and proteins but also as a valuable tools in understanding biological aging. In fact, *S. cerevisiae* has a long history as the workhorse of pharmaceutical discovery research [17]. The uses of yeast explained in this chapter are including primary pathways involved in the mechanism of yeast aging belong to pro- or anti-aging mechanisms, and sections on the use of yeast for elucidating anti-aging compounds through the popular methods that was already utilized by previuos researcher's. To this end, we would state that there are still limitation in regard with anti-aging study or review of using yeast as a model organism for elucidating anti-aging compounds in our country, Indonesia. At the same time, this country has an abundant sources of medicinal compounds derived from terestrial to aquatic regions. Therefore, we hope that this book chapter could provide new alternatively insight for pharamaceutical study focusing to discover active compounds with anti-aging properties.

## 2. Aging interventing mechanisms in *S. cerevisiae*

Biological aging in *S. cerevisiae* involved complicated mechanisms including, cellular, physiological and molecular as well as intervented by environmental growth conditions. Interestingly, almost all of yeast model, particularly *S. cerevisiae* has sophisticated response while growth in Calorie Restriction (CR) condition, which scientist's oftenly used those phenomenon to learn aging pathway in yeast model. CR, a reduction in nutrient availability without malnutrition, is known to expand lifespan in a wide range of organisms from yeast to primates [11, 18]. Of note, deeply effort has been devoted to understanding the pathways that mediate the benefits of dietary restriction, since interventions that target these pathways may be effective in humans against the diseases of aging. In the yeast anti-aging assay, dietary restriction is usually generated by reducing glucose concentration from 2% to 0.5 or 0.05% in the growth culture [6], as well as restriction for amino acids has also been reported to exert lifespan [19].

Ultimately, CR reported could affect in some distinctive pathways in yeast cells. The first pathway is nutrient-sensing which reduced activity of two major nutrient sensing pathways, due to CR condition, could extend yeast life span. Both nutrient sensing pathways are focused on an amino acid-sensing pathway, including the serine–threonine kinase SCH9 and the target of rapamycin (TOR). Notably, deletion or inhibition of SCH9 and TOR causes an increase of up to several fold in yeast life span. Alterations to reduce nutrient and protein synthesis in CR condition are strongly implicated in extension of yeast lifespan by reduced TOR/SCH9. Extension of yeast lifespan by reduced activity of the TOR pathway depends on the transcription factor Gis1, which activates many protective systems including Mn-SOD [6, 20].

Further, the second pathway includes three proteins including adenylate cyclase (AC), protein kinase A (PKA), and Ras which will inhibit by CR conditions. The activation of two transcription factors (Msn2 and Msn4) that control cellular protection systems is required to mediate the effect of reduced Ras-AC-PKA signaling on yeast lifespan extension. Extension of yeast lifespan by these pathways needs the antioxidant enzyme Mn-SOD (superoxide dismutase), which scavenges the superoxide free radical [11, 21]. Intriguingly, superoxide level increases during yeast aging and is reduced in yeast mutants deficient in Tor-SCH9 or Ras-AC-PKA signaling. As for the yeast cells grow in the high glucose medium, could produce ethanol or acetic acid, which also contribute to chronological aging. Interestingly, deletion of SCH9 or TOR1 promotes removal of ethanol and acetic acid and accumulation of glycerol in the medium and further extend chronological life span by mechanisms similar to those of dietary restriction [22]. More importantly, decreased signaling by the Tor-SCH9 and Ras-AC-PKA pathways is important in response to glucose restriction as well as increased transcriptional activity of Msn2 and Msn4, and the consequent affecting the expression of Pnc1 [nicotinamide deaminase that promotes the activity of the nicotinamide adenine dinucleotide (NAD)- dependent deacetylase Sirtuin 2/Sir2]. The Sir2 have been extensively studied for their potential role as conserved modulators of anti-aging in a various of organisms, including mammals [23]. One mechanism by which Sir2 activity promotes yeast longevity is by suppressing homologous recombination in the rDNA that can promote the formation of extrachromosomal rDNA circles (ERCs). In fact, rDNA instability in general suggested the primary defect causing senescence and cell death [24].

Notably, another crucial mechanism that closely related with yeast aging is autophagy. This cellular process is reported as a highly conserved in organisms from yeast to human, which involves degradation of damaged organelles, circulation of amino acids, proteins, and other metabolites. It also regulates the genomic integrity via suppression of cell division in yeast under CR condition. Notably, decreased or dysfunction expression of autophagy genes leads to shorter lifespan in yeast and fruit fly. Conversely, enhanced autophagy promotes the longevity in aging models and suggested could protect against aging and age-related disorders [25, 26]. Another mechanisms which closely related with yeast aging is mitochondrial

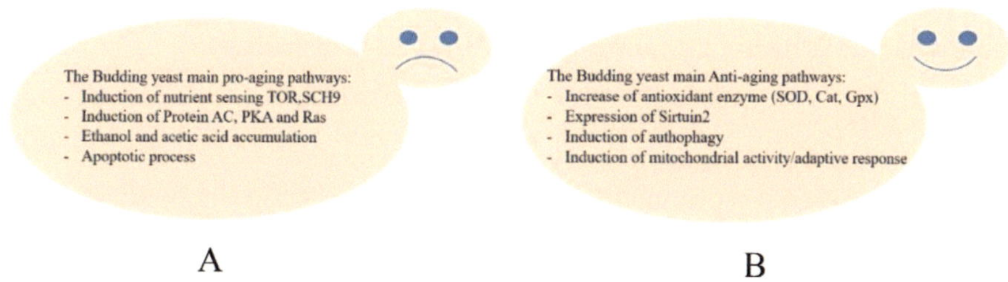


**Figure 2.**
*(A) Some main pro-aging and (B) anti-aging pathways in* S. cerevisiae. *(TOR: target of rapamycin, SCH9: serine–threonine kinase, AC: Adenylate cyclase, PKA: Protein kinase A, SOD: Superoxide dismutase, Cat: Catalase, Gpx: Glutathione peroxidase).*

adaptive response signaling. Mitochondrial organelle is known to have a basic role in aging and age-related diseases. This organelle contributes to the ATP production, cell homeostasis, and imbalanced reactive oxygen species (ROS) creating a basic role of the cells regulation [27]. In common condition, mitochondrial produces toxic ROS as by product of respiration process. However, on the CR condition, mitochondrial would be more active due to the shift metabolic process occurs from fermentation to respiration (CR) resulting ROS at the initial growth phases, as well. Consequently, yeast cells will adapt (pre-adaptation) with the ROS molecules strating from early growth stage and thus might activate defence mechanisms in the late of growth phase ensuring protection against higher doses of ROS. Those defence mechanisms likely activated antioxidative enzyme i.e. superoxide dismutase (SOD), catalase or glutathione peroxidase and therefore increase yeast lifespan (**Figure 2**) [28, 29].

Other than the above mechanisms in relation with CR condition resulting yeast lifespan extension. CR was also reported as having other substantial effects in yeast cells i.e. apoptotic process that accelerate aging process, repairing protein damage, NAD+ homeostasis, vacuolar function, genome stability, ribosom biogenesis, proteolysis regulation, and cell hyperthrophy [30]. These valuable insights reflect that CR condition in yeast cells could modulate numerous pathways affected in biological aging mechanism. As for forefront anti-aging methods strategy, yeast cells growth on CR oftenly use as for positive control, whereas growing yeast cells on the normal growth medium with 2% glucose utilize as treatment for anti-aging compounds screening. If the yeast cells viability derived from compounds treatment has similar or higher than positive control, it suggested that corresponding compounds have anti-aging activity in yeast cells. Further research usually applied to investigate the precise mechanisms which modulated by those promising anti-aging compounds inside yeast cells.

## 3. Methods for investigating anti-aging activity in *S. cerevisiae*

On the basis of the current literatures, there were established various potential methods for investigating anti-aging activity derived from chemical compound by using *S. cerevisiae* as a model organism. It is including spotting test, Chronological Life Span (CLS) using Total Plate Count (TPC) analysis, Replicative Life Span (RLS) assay using microscopic observation, Propidium Iodida – Flow Cytometry Analysis (PI-FCA) analysis (see detail in Ocampo and Barrientos, 2011), Bac-light method (see detail in [31] Zhang and Fang, 2004), staining with methylen blue or phloxin B, physiological assay using luceferin reaction or rhodamine B (see detail in [32], and High-Throughput Rapid Chronological Lifespan (HTRCL) based on MTT assay (See detail in [33]). However, in this section we would highlight only the most 3 popular methods including spotting test, CLS, and RLS analysis.

### 3.1 Spotting test

This particular method is commonly used in order to observe the viability of *S. cerevisiae* cells through the phenotype spotting shape on the surface of solid growth medium. In fact, this method is also applied not only for *S. cerevisiae*, but also for other yeast to assay anti-aging activity of compounds, including *Schizosaccharomyces pombe*, *Kluyveromyces lactis*, *Candida albicans*, and *Cryptococcus neoformans* [7]. In addition, microbiologist oftenly utilize this method to search the specific microbial mutant that sensitive to particular conditions existing in the solid medium i.e. oxidative stress, antibiotic resistance, heat, osmotic, nutrient limitation, etc. Briefly, spot test was conducted by inoculating the fresh logarithmic *S. cerevisae* cells to the yeast liquid medium in the test tube with an appropriate initial OD600 of 0.05–0.1. Subsequently, the corresponding compound supplemented to the same yeast liquid medium with a certain concentration, and culture were incubated until reached the stationary phase (20–25 days) at the optimal growth temperature. Periodically, yeast treatment culture (usually at day 5, 10, 15, and 20) spotted to the yeast solid culture. As for the spotting test was initially performs by serially dilution of the treatment culture at the selected incubation time until $10^{-4}$ dilution value, then spotting applied from each dilution to the yeast solid medium. Further, the results are incubated for 3 days at the optimal growth temperature prior observed. The density of the grown cells from each spot is considered as the viability of *S. cerevisiae* cells (**Figure 3**) (see detail in [1, 34]).

Spotting test was reported as having some advantages in relation to assay anti-aging compounds including simple, fast handling and relatively low cost compare than CLS or other methods. However, this method could only represent a qualitative result of yeast cells viability through the yeast cells density, and thus it should be supported by other methods. As long as for the preliminary screening of numerous compounds acting as anti-aging, spotting method will be recommended. Some previous studies reported the usefulness of spot test to examine the particular compound for delaying aging in *S. cerevisiae*. Xiang et al., [35] was succeded to promote anti-aging mechanism in this yeast using phloridzin (an apple polyphenol) treatment. Apple extract, artemisinin, and roselle petal extract were also reported could increase *S. cerevisiae* lifespan using spotting test method [36–38]. Along with that, *S. cerevisiae* was succesfully used for obtaining the anti-aging activity of clove bud extract and some medicinal plant extract [1, 39].

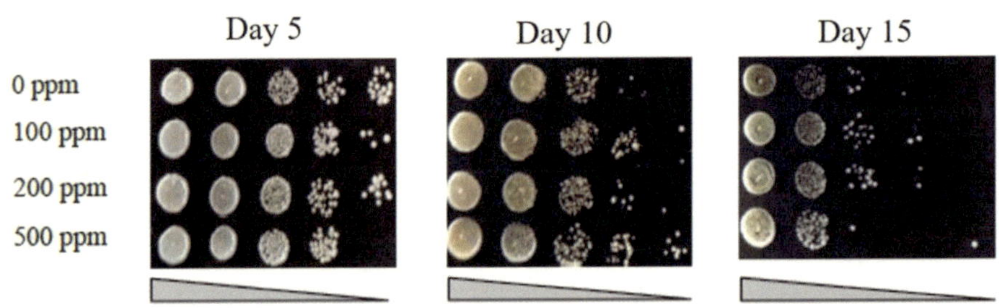


**Figure 3.**
*Result of spotting test from clove bud extract using* S. cerevisiae *as model organism. Cells grown in yeast culture without extract addition (0 ppm) is designed as a control. The density of the grown cells from each spot considered as the viability of* S. cerevisiae *cells. Spotted at day 15 after treatment with 100 and 200 ppm compound showed higher density than 0 ppm (control treatment), indicating anti-aging activity of corresponding compound to exert the yeast lifespan (picture permitted from [1], originally published in IOP Conf. Series: Earth and Environmental Science. doi: 10.1088/1755-1315/299/1/012059).*

### 3.2 CLS method using TPC analysis

CLS defines to the length of time a non-dividing cell can maintain viability, as refers to the its ability to re-enter the cell cycle process after a prolonged period of quiescence. Thus, CLS has been exhibited as a model of the viability of post-mitotic yeast cells [12]. Traditionally, CLS has been examined by culturing fresh logarithmic yeast cells on a particular flask until reached the stationary phase (20–25 days) in liquid culture with an appropriate initial OD600 of 0.05–0.1. As for compound treatment is applied to the liquid culture medium soon thereafter yeast innoculated. Further, the yeast cell survival is measured as a function of time by dilution and plating onto a nutrient-rich agar medium at the periodic time (i.e each 3 days). Subsequently, viability is then calculated on the basis of the number of colonies (colony forming units: CFUs) on the nutrient plate agar arising [40, 41]. The budding yeast *S. cerevisiae* reported entering a stationary/death phase after 20–25 days incubation [6]. Therefore, if compound treatment could prolong the yeast lifespan beyond 25 days, it suggested as the promising anti-aging materials. Of note, CLS assay is widely applied not only for *S. cerevisiae* but also for other yeast i.e. *S. pombe* or *K. lactis* for observing the bioactivity of prospective anti-aging compounds (i.e **Figure 4**, CLS applying in *S. pombe*).

Ultimately, CLS method requires a relatively large investment of materials, investigator time and belong to laborious, therefore is not suited for high-throughput screening anti-aging compounds. However, through this particular method, it was obtained the quantitatively results and thus could provide deeply insight for representing the cell viability of the yeast cells. Numerous studies has been applied CLS method for assaying anti-aging compounds derived from multi-resources i.e. Nakaya et al., [42] assayed Beauveriolide I isolated from mushroom or Sunthonkun *et al*., [40] examined anti-aging of pigmented rice. In addition, other compounds including Dimethylchalcone, Cucurbitacin B, and Astaxanthine was exhibited anti-aging activity in *S. cerevisiae* through CLS analysis [14–16].

### 3.3 RLS method

RLS assay is simple conceptually and shows an advantage of the fact that *S. cerevisiae* cells divide by asymmetric budding, with the daughter cells that is produced

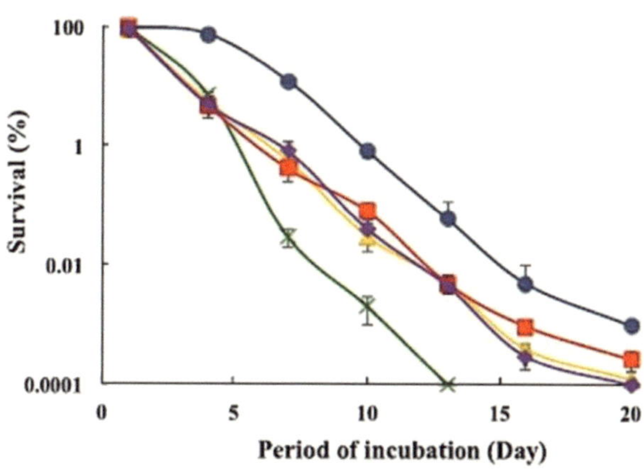


**Figure 4.**
*CLS result using* S. pombe *(green line as a control treatment) after treatment with some compounds (red, orange, purple and yellow lines showed higher viability day than control treatment). In* S.pombe, *the stationary phase will reach after 13–15 days incubation at the optimal temperature (Prastya et al., 2020).*

being smaller than the mother cells of which it is derived. Daughter cells were isolated on a solid growth media and, once they started dividing, all daughter cells were removed. Longo (2012) founded that an individual cells do not divide forever, instead they would stop after a limited number of cells divisions (around of 20–25) and going to a short post-replicative state followed by death cells. The RLS method determines the number of daughters of a single mother cell, which can asexually produce prior to senescence. The mother-daughter cell asymmetry in *S. cerevisiae* cells can be easily observed under the light microscope, allowing the development of the RLS assay [6, 43].

As for RLS assay, the *S. cerevisiae* cells inoculated onto 5 mL galactose or glucose liquid medium and further incubated in a shaking incubator for 48 h at the optimal temperature. Subsequently, 1 mL of yeast cells culture was centrifuged and pellet washed with distilled water or phosphate buffer solution (PBS). After counting using a hemocytometer, 4000–5000 cells are plated on agar plates medium containing chemical compounds need to be assay for antiaging activity, and plates were further incubated for 2 days at the optimal temperature. The 40 microcolonies that formed on the agar plates were randomly observed under a microscope, and daughter cells were counted. Recently, *S. cerevisase* yeast cells reported as the most powerfull model organism for RLS analysis [44, 45]. The results of RLS assay usually show in the particular graphic that indicating the viability number from each generations of the yeast cells. If the viability of the yeast cells after compound treatment is higher than control treatment, it is indicating the potential anti-aging properties.

RLS method is reported as having some major weakness which it makes less effective for high-throughput approaches. It is including time-consuming, laborious and relatively intricate in technique due to applying microscopic cells observation prior for plating in the plate medium during assays. Nevertheless, this particular methods will devote precisely quntitatively results and thus could represent the number of yeast cells generation between control and anti-aging compound treatments. To date, some previous studies were reported for using *S. cerevisiae* RLS method to examine anti-aging compound derived from various natural products, including Ganodermasides isolated from mushroom, or Hesperidin from citrus [46, 47]. Current studies were also informed anti-aging assay of some compounds i. e Parishin and Cucurbitacin B using *S. cerevisiae* RLS method, as well [15, 45].

## 4. Conclusions

Antiaging study in yeast was popular using CR condition which has numerous response to prolong yeast lifespan. Aging pathway in CR belong to pro-and anti-aging pathways. As for pro-aging are including TOR, SCH9, Ras protein, AC, PKA, ethanol accumulation, and apoptotic process. On the other hand, anti-aging pathways are including induction of antioxidative enzymes, sirtuin2, autophagy and adaptive response thorough mitochondrial adaptive ROS signaling. CR condition usually use for positive control, while treatment conducted in non-CR/high 2% glucose medium. There are numerous methods for anti-aging study, which the most popular is spotting test, CLS and RLS assays.

Recently, anti-aging in *S. cerevisiae* research was developed sophisticated, derived from previous sources including natural compound derived from terestrial or aquatic organism. In fact, other sources also potentially developed i.e. semisynthetic or synthetic compound as the preliminary screening for anti-aging compounds.

Even the data from studies could be revisited and mined for potential bioactive substances, data obtained in yeast should not be over-interpreted unduly, and when aiming for applications in humans, validation of compounds in multicellular organisms should be done. So far, the potential of yeast to unravel novel pharmacological interventions against aging is far-reaching, however that it will continue to contribute substantially not only to drug discovery but also in other field such as fermented food, biochemical and bioenergy production.

## Acknowledgements

The authors would like to thank Dr. Rika Indri Astuti Department of Biology, IPB University, Indonesia for permitting the pictures.

## Conflict of interest

On behalf of all authors, we declare no conflict of interest.

## Author details

Yanni Sudiyani*, Muhammad Eka Prastya, Roni Maryana, Eka Triwahyuni and Muryanto
Research Center for Chemistry, Indonesian Institute of Sciences, Serpong Tangerang Selatan, Banten, Indonesia

*Address all correspondence to: sudiyani@gmail.com

## References

[1] R. I. Astuti, S. Listyowati, and W. T. Wahyuni, "Life span extension of model yeast *Saccharomyces cerevisiae* upon ethanol derived-clover bud extract treatment," in *IOP Conference Series: Earth and Environmental Science*, 2019, vol. 299, no. 1, p. 12059.

[2] G. M. Walker and N. A. White, "Introduction to Fungal Physiology," *Fungi*. pp. 1-35, Nov. 22, 2017, doi: doi:10.1002/9781119374312.ch1.

[3] A. Goffeau *et al.*, "Life with 6000 genes," Science (80-. )., vol. 274, no. 5287, pp. 546-567, 1996.

[4] R. Montes de Oca *et al.*, "Yeast: description and structure," *Yeast Addit. Anim. Prod. Tamilnadu, PubBioMed Cent. Res. Publ. Serv.*, pp. 4-13, 2016.

[5] M. H. Barros, F. M. da Cunha, G. A. Oliveira, E. B. Tahara, and A. J. Kowaltowski, "Yeast as a model to study mitochondrial mechanisms in ageing," Mech. Ageing Dev., vol. 131, no. 7-8, pp. 494-502, 2010.

[6] V. D. Longo, G. S. Shadel, M. Kaeberlein, and B. Kennedy, "Replicative and chronological aging in *Saccharomyces cerevisiae*," Cell Metab., vol. 16, no. 1, pp. 18-31, 2012.

[7] A. Zimmermann, S. Hofer, T. Pendl, K. Kainz, F. Madeo, and D. Carmona-Gutierrez, "Yeast as a tool to identify anti-aging compounds," *FEMS Yeast Res.*, vol. 18, no. 6, Sep. 2018, doi: 10.1093/femsyr/foy020.

[8] A. Olsen, M. C. Vantipalli, and G. J. Lithgow, "Using Caenorhabditis elegans as a model for aging and age-related diseases," Ann. N. Y. Acad. Sci., vol. 1067, no. 1, pp. 120-128, 2006.

[9] S. L. Helfand and B. Rogina, "Genetics of aging in the fruit fly, Drosophila melanogaster," Annu. Rev. Genet., vol. 37, no. 1, pp. 329-348, 2003.

[10] M. Kaeberlein, C. R. Burtner, and B. K. Kennedy, "Recent developments in yeast aging," *PLoS Genet*, vol. 3, no. 5, p. e84, 2007.

[11] L. Fontana, L. Partridge, and V. D. Longo, "Extending Healthy Life Span—From Yeast to Humans," *Science (80-. ).*, vol. 328, no. 5976, pp. 321 LP – 326, Apr. 2010, doi: 10.1126/science.1172539.

[12] C. J. Murakami, C. R. Burtner, B. K. Kennedy, and M. Kaeberlein, "A method for high-throughput quantitative analysis of yeast chronological life span," Journals Gerontol. Ser. A Biol. Sci. Med. Sci., vol. 63, no. 2, pp. 113-121, 2008.

[13] M. Kaeberlein, "Lessons on longevity from budding yeast," Nature, vol. 464, no. 7288, pp. 513-519, 2010.

[14] D. Carmona-Gutierrez *et al.*, "The flavonoid 4, 4′-dimethoxychalcone promotes autophagy-dependent longevity across species," Nat. Commun., vol. 10, no. 1, pp. 1-17, 2019.

[15] Y. Lin *et al.*, "Cucurbitacin B Exerts Antiaging Effects in Yeast by Regulating Autophagy and Oxidative Stress," *Oxid. Med. Cell. Longev.*, vol. 2019, 2019.

[16] S. SJ, B. Veerabhadrappa, S. Subramaniyan, and M. Dyavaiah, "Astaxanthin enhances the longevity of *Saccharomyces cerevisiae* by decreasing oxidative stress and apoptosis," *FEMS Yeast Res.*, vol. 19, no. 1, Jan. 2019, doi: 10.1093/femsyr/foy113.

[17] P. Ross-Macdonald, "Growing Yeast for Fun and Profit: Use of *Saccharomyces cerevisiae* as a Model System in Drug Discovery," Model Org. Drug Discov., pp. 9-39, 2003.

[18] R. M. Anderson and R. Weindruch, "The caloric restriction paradigm: implications for healthy human aging," Am. J. Hum. Biol., vol. 24, no. 2, pp. 101-106, 2012.

[19] J. C. Jiang, E. Jaruga, M. V Repnevskaya, and S. M. Jazwinski, "An intervention resembling caloric restriction prolongs life span and retards aging in yeast," FASEB J., vol. 14, no. 14, pp. 2135-2137, 2000.

[20] M. Kaeberlein *et al.*, "Regulation of yeast replicative life span by TOR and Sch9 in response to nutrients," Science (80-. )., vol. 310, no. 5751, pp. 1193-1196, 2005.

[21] P. Fabrizio, F. Pozza, S. D. Pletcher, C. M. Gendron, and V. D. Longo, "Regulation of longevity and stress resistance by Sch9 in yeast," Science (80-. )., vol. 292, no. 5515, pp. 288-290, 2001.

[22] P. Fabrizio *et al.*, "Sir2 blocks extreme life-span extension," Cell, vol. 123, no. 4, pp. 655-667, 2005.

[23] T. Finkel, C.-X. Deng, and R. Mostoslavsky, "Recent progress in the biology and physiology of sirtuins," Nature, vol. 460, no. 7255, pp. 587-591, 2009.

[24] D. L. Lindstrom, C. K. Leverich, K. A. Henderson, and D. E. Gottschling, "Replicative age induces mitotic recombination in the ribosomal RNA gene cluster of *Saccharomyces cerevisiae*," *PLoS Genet*, vol. 7, no. 3, p. e1002015, 2011.

[25] D. C. Rubinsztein, G. Mariño, and G. Kroemer, "Autophagy and aging," Cell, vol. 146, no. 5, pp. 682-695, 2011.

[26] J. K. Tyler and J. E. Johnson, "The role of autophagy in the regulation of yeast life span," Ann. N. Y. Acad. Sci., vol. 1418, no. 1, pp. 31-43, Apr. 2018, doi: 10.1111/nyas.13549.

[27] M. T. Lin and M. F. Beal, "Mitochondrial dysfunction and oxidative stress in neurodegenerative diseases," Nature, vol. 443, no. 7113, pp. 787-795, 2006.

[28] Y. Pan, E. A. Schroeder, A. Ocampo, A. Barrientos, and G. S. Shadel, "Regulation of yeast chronological life span by TORC1 via adaptive mitochondrial ROS signaling," Cell Metab., vol. 13, no. 6, pp. 668-678, 2011.

[29] G. S. Shadel, "Live longer on MARS: a yeast paradigm of mitochondrial adaptive ROS signaling in aging," *Microb. cell (Graz, Austria)*, vol. 1, no. 5, pp. 140-144, Apr. 2014, doi: 10.15698/mic2014.05.143.

[30] P. Fabrizio and V. D. Longo, "The chronological life span of *Saccharomyces cerevisiae*," Aging Cell, vol. 2, no. 2, pp. 73-81, 2003.

[31] T. Zhang and H. H. P. Fang, "Quantification of *Saccharomyces cerevisiae* viability using BacLight," Biotechnol. Lett., vol. 26, no. 12, pp. 989-992, 2004, doi: 10.1023/B:BILE.0000030045.16713.19.

[32] M. Kwolek-Mirek and R. Zadrag-Tecza, "Comparison of methods used for assessing the viability and vitality of yeast cells," FEMS Yeast Res., vol. 14, no. 7, pp. 1068-1079, 2014.

[33] Z. R. Belak, T. Harkness, and C. H. Eskiw, "A rapid, high-throughput method for determining chronological lifespan in budding yeast," *J. Biol. Methods*, vol. 5, no. 4, 2018.

[34] M. E. Prastya, R. I. Astuti, I. Batubara, and A. T. Wahyudi, "Bacillus sp. SAB E-41-derived extract shows antiaging properties via ctt1-mediated oxidative stress tolerance response in yeast Schizosaccharomyces pombe," *Asian Pac. J. Trop. Biomed.*, vol. 8, no. 11, p. 533, 2018.

[35] L. XIANG *et al.*, "Anti-Aging Effects of Phloridzin, an Apple Polyphenol, on Yeast via the SOD and Sir2 Genes," Biosci. Biotechnol. Biochem., vol. 75, no. 5, pp. 854-858, 2011, doi: 10.1271/bbb.100774.

[36] V. Palermo, F. Mattivi, R. Silvestri, G. La Regina, C. Falcone, and C. Mazzoni, "Apple can act as anti-aging on yeast cells," *Oxid. Med. Cell. Longev.*, vol. 2012, 2012.

[37] D. Wang, M. Wu, S. Li, Q. Gao, and Q. Zeng, "Artemisinin mimics calorie restriction to extend yeast lifespan via a dual-phase mode: a conclusion drawn from global transcriptome profiling," Sci. China Life Sci., vol. 58, no. 5, pp. 451-465, 2015, doi: 10.1007/s11427-014-4736-9.

[38] A. R. I. Sarima and A. Meryandini, "Modulation of aging in yeast *Saccharomyces cerevisiae* by roselle petal extract (Hibiscus sabdariffa L.)," Am. J. Biochem. Biotechnol., vol. 15, no. 1, pp. 23-32, 2019.

[39] P. Dakik *et al.*, "Discovery of fifteen new geroprotective plant extracts and identification of cellular processes they affect to prolong the chronological lifespan of budding yeast," *Oncotarget*, vol. 11, no. 23, p. 2182, 2020.

[40] P. Sunthonkun, R. Palajai, P. Somboon, C. L. Suan, M. Ungsurangsri, and N. Soontorngun, "Life-span extension by pigmented rice bran in the model yeast *Saccharomyces cerevisiae*," *Sci. Rep.*, vol. 9, no. 1, p. 18061, 2019, doi: 10.1038/s41598-019-54448-9.

[41] M. E. Prastya, R. I. Astuti, I. Batubara, H. Takagi, and A. T. Wahyudi, "Natural extract and its fractions isolated from the marine bacterium Pseudoalteromonas flavipulchra STILL-33 have antioxidant and antiaging activities in Schizosaccharomyces pombe," *FEMS Yeast Res.*, vol. 20, no. 3, p. foaa014, 2020.

[42] S. Nakaya, S. Mizuno, H. Ishigami, Y. Yamakawa, H. Kawagishi, and T. Ushimaru, "New rapid screening method for anti-aging compounds using budding yeast and identification of beauveriolide I as a potent active compound," Biosci. Biotechnol. Biochem., vol. 76, no. 6, pp. 1226-1228, 2012.

[43] R. K. Mortimer and J. R. Johnston, "Life span of individual yeast cells," Nature, vol. 183, no. 4677, pp. 1751-1752, 1959.

[44] S.-J. Lin and N. Austriaco, "Aging and cell death in the other yeasts, Schizosaccharomyces pombe and Candida albicans," FEMS Yeast Res., vol. 14, no. 1, pp. 119-135, 2014.

[45] Y. Lin, Y. Sun, Y. Weng, A. Matsuura, L. Xiang, and J. Qi, "Parishin from Gastrodia elata extends the lifespan of yeast via regulation of Sir2/Uth1/TOR signaling pathway," *Oxid. Med. Cell. Longev.*, vol. 2016, 2016.

[46] Y. WENG *et al.*, "Ganodermasides C and D, Two New Anti-Aging Ergosterols from Spores of the Medicinal Mushroom Ganoderma lucidum," Biosci. Biotechnol. Biochem., vol. 75, no. 4, pp. 800-803, 2011, doi: 10.1271/bbb.100918.

[47] K. SUN, L. XIANG, S. ISHIHARA, A. MATSUURA, Y. SAKAGAMI, and J. QI, "Anti-Aging Effects of Hesperidin on *Saccharomyces cerevisiae* via Inhibition of Reactive Oxygen Species and UTH1 Gene Expression," *Biosci. Biotechnol. Biochem.*, vol. advpub, p. 1202232809, 2012, doi: 10.1271/bbb.110535.

Chapter 7

# Secondary Metabolites from *Saccharomyces cerevisiae* Species with Anticancer Potential

*Muhammad Jahangeer, Areej Riasat, Zahed Mahmood, Muhammad Numan, Naveed Munir, Mehvish Ashiq, Muhammad Asad, Usman Ali and Mahwish Salman*

## Abstract

Chemotherapeutic agents produce from numerous sources such as animals, plants and micro-organisms are derived from the natural products. Although the existing therapeutic pipeline lacks fungal-derived metabolites, but hundreds of secondary metabolites derived from fungi are known to be possible chemotherapies. Over the past three decades, several secondary metabolites such as flavonoids, alkaloids, phenolic and polyketides have been developed by *Saccharomyces cerevisiae* species with exciting activities that considered valued for the growth of new chemotherapeutic agents. Many secondary metabolites are protective compounds which prevent abiotic and biotic stresses, i.e. predation, infection, drought and ultraviolet. Though not taking part in a living cell's central metabolism, secondary metabolites play an important role in the function of an organism. Nevertheless, due to slow biomass build-up and inadequate synthesis by the natural host the yield of secondary metabolites is low by direct isolation. A detailed comprehension of biosynthetic pathways for development of secondary metabolites are necessary for *S. cerevisiae* biotransformation. These metabolites have higher inhibitory effect, specificity among cancer and normal cells, and the mechanism of non-apoptotic cell killing. This study shows the significance of bioactive compounds produced by *S. cerevisiae* species with their possible activity and value in chemotherapeutic drugs pipeline. The isolation and alteration of these natural secondary metabolites would promote the development of chemotherapeutic drugs.

**Keywords:** *Saccharomyces cerevisiae*, secondary metabolites, anticancer activity, synthetic biology, bioactive compounds

## 1. Introduction

Yeast is a single-celled eukaryotic microorganism that belongs to the kingdom of fungi. About 1500 yeast species have been correctly described since the discovery of the first yeast [1]. Yet 1% of all known fungal members are stated to be yeast species. *Saccharomyces cerevisiae*, budding yeas, also represents a typical industrial microorganism used in basic molecular biology research as a main model organism and was the first eukaryotic organism to have completely sequenced its genome.

A single-cell fungus, *Saccharomyces cerevisiae*, is also known as a Baker's yeast [2]. *Saccharomyces cerevisiae* is the eukaryotic microorganism most extensively studied, which allows us to understand the eukaryotic cell biology and subsequently the physiology of human. For several hundred years, in food processing and alcoholic drinks, *S. cerevisiae* have been used, and this organism is still used today in a variety of different pharmaceutical processes. It functions very well as it is non-pathogenic and is classified by GRAS organism (generally regarded as safe) due to the long history of use in the development of consumables such as ethanol and baker's yeast [3].

The sum of all organismic biochemical reactions can be described as metabolism. Metabolites are the products and intermediates of metabolism and are generally limited to small molecules. Natural products come from a variety of sources, including animal species, land-based plants, aquatic organisms, land invertebrates and vertebrates, microorganisms as secondary metabolite products [4]. The word "secondary" proposed by A. Kossel in 1891 means that while in any living cell primary metabolites are present, the secondary metabolites are only present at a by-product and are of no significant importance to the life of the organism. Although the primary metabolism derives secondary metabolites, these secondary metabolites do not constitute the organism's fundamental molecular framework. Although the secondary metabolites do not participate in central metabolic processes of a living cell, they play an important role in the function of an organism. Many secondary bioactive compounds defend against biotic and abiotic and biotic stresses including predation, cancer, drought and ultraviolet radiations. Its absence does not reduce the life of an organism, which is a feature contrary to primary metabolite, instantly, but largely affected the organism' survival. Currently there are a range of analytical platforms for metabolomics research, among which are including mass spectrometry direct infusion (MS), gas chromatography liked to mass spectrometry (GC– MS), gas chromatography linking to mass spectrometry two-dimensional (GC, GC– MS), liquid chromatography coupled to mass spectrometry (LC–MS), capillary electrophoresis to mass spectrometry coupling (CE–MS), and proton nuclear magnetic resonance spectroscopy (1H NMR) [5].

These complex structural and chemical molecules serve as an excellent therapeutic class to cure many diseases. Around 80 percent of all drugs were derived from plant sources at the beginning of 1900. Alexander Fleming's discovery of penicillin from Penicillium notatum in 1928 marked a major change as a source of natural product from plants to micro-organisms. In medicine, agriculture, food industry and scientific research microorganism-derived compounds since then have been used [6]. But, due to the slow accumulation by biomass and inadequate synthesis by the host, the production of secondary metabolites by direct isolation is poor. Exhaustive awareness of biosynthetic pathways for the production of secondary metabolites are necessary for the development of *S. cerevisiae* biotransformation. Progress in synthetic biology has made it possible to develop a number of bioinformatics tools that can be utilized to create new regulatory elements and secondary metabolite synthesis pathways [7].

Because of its similar metabolism *Saccharomyces cerevisiae* was used as a biologically similar model for higher eukaryote organisms. The expression of heterogeneous pathways is less difficult relative to other kinds because of its eukaryotic nature. For example, prokaryotes do not conduct any post-translational modifications, and protein miss-folding and membrane translation can be an annoyance [8]. During preclinical or clinical testing, a well-characterized yeast metabolite, such as flavonoids, alkaloids, phenolic and polyketides, exhibit remarkable anti-tumor properties unexpectedly. Although its basic action mechanism is still being studied, evidence shows that its actions are operationally directed toward core regulatory pathways and dysregulated enzymes during cancer pathogenesis.

Since ancient time, from the days of the Pharaohs in ancient Egypt, cancer continues to plague humanity. It derives its name from Hippocrates, father of medicine, who had been using the Greek-named "Karkinos" to talk about tumors, nevertheless these earliest view about this infection are different from modern concept. Cancer has been characterized as an irregular growth in cells caused by several changes in gene-expression leading to dysregulated equilibrium between death of cell and proliferation eventually developing into cell populations that can invade and metastasize tissues in distant sites and cause severe host death if left unsanctioned according to the World Health Organization (WHO). About 60 percent of all anticancer medications currently available in clinics are natural or derived from natural product modification substances. In the 1950s interest in the discovery and production of vinblastine, vinka alkaloids and vincristine, and later taxol from Pacific yew bark, *Taxus brevifolia*, in the pursuit of natural anticancer agents began earnestly. Then, our awareness of the metabolites of *S. cerevisiae* have also been further progressed in cancer therapy and have reported antitumor action against the overwhelming majority of cancers, like lymphoma, leukemia and solid tumors. Far from the suppression of tumor development, the delay of tumor progression, and an effect on tumor-cell metastatic and invading therapy, these metabolite combinations have almost all demonstrated strong therapeutic benefit at the preclinical level. Some of these allegedly promising metabolite compounds are discussed individually below [9].

## 2. Production of secondary metabolites via *S. cerevisiae*

Apart from essential metabolites (carbons, proteins, amino acids, vitamins, acetones, ethanol, etc.), *S. cerevisiae* offers a wide range of secondary metabolites during active cell development, including toxins, antibiotics, fatty acids, alkaloids, alcohols, ketones etc. Secondary metabolites (SM's), which are not necessary for organism growth, are classified as diverse low molecular-weighted compounds. For a number of purposes, *S. cerevisiae* use secondary metabolites such as stress prevention, predation defense, competitiveness, communications, pathogenicity and exposure of other organisms. A small number of primary metabolism precursor metabolites are used to biosynthesize secondary metabolites [10].

A number of bioactive compounds, including terpenoids, polyketides, alkaloids and non-ribosomal peptides can be produced in *Saccharomyces cerevisiae* (**Figure 1**). Such precursors are mainly short chain carboxylic acids (for example, acetyl-coA) or amino acids in *Saccharomyces cerevisiae* species which are connected to synthesized polyketides by backbone enzymes such as polyketide synthases (PKSs). Terpenoids production is catalyzed by terpene cyclase and production starting from dimethyl allyl diphosphate derived from (or isoprene). Non-ribosomal peptide synthetases (NRPSs) using modified and natural amino acids to synthesize non-ribosomal peptides. Diverse enzymes are used to catalyze the synthesis of alkaloids from amino acid. These secondary metabolites can naturally be synthesized as industrial products, particularly pharmaceutical products, from host (native) cells.

Many other reports have shown that the genes involved in biosynthetic secondary metabolites are known as biosynthetic gene clusters (BGCs). This have been shown that genes that encode significant biosynthesis (for example; polyketides synthase) are supplementary enzymes and precursors in the pathways of biosynthesis. Consequently, all significant genes are discovered in the BGCs that participate in the synthesis of bioactive compounds [11, 12]. The production and complementary regulation of catalytic properties can produce incredibly useful secondary metabolites for these biochemical transformations BGC enzymes [13].

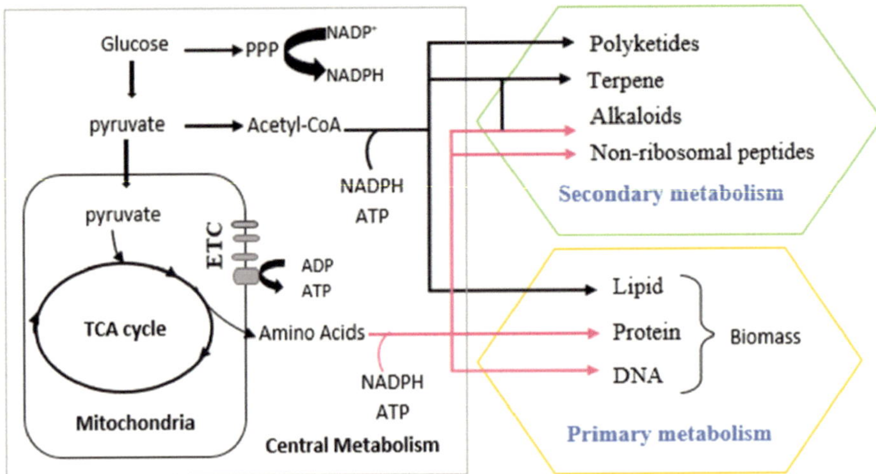


**Figure 1.**
*A short description of the formation pathways of secondary metabolites [11, 12]. Four sub-groups of microorganism and plant-generative metabolites involve terpenoids, polyketides, non-ribosomal peptides and alkaloids. Biosynthetic genes (BGC's) are structurally classified into the genes that encrypt major biosynthetic enzymes, supplementary enzymes, precursors in the process of biosynthesizing, resistance factors or transporters in secondary metabolite formation.*

The methods and technology used to improve pharmaceutical production of secondary metabolites in the *S. cerevisiae* have advanced rapidly in recent years.

We discuss below the progress in developing biosynthetic pathways of *Saccharomyces cerevisiae*' secondary metabolites with anticancer potential: alkaloids, phenolic, terpenoids, polyketides, non-ribosomal peptides, and vitamin C.

## 3. Alkaloids

Several studies have used *S. cerevisiae* in recent years as a host to engineer the biosynthetic process of alkaloids. Alkaloids are complex nitrogen molecules that are extremely bioactive. There are reportedly approximately 50 alkaloid medications, including vincristine (cancer drug), codeine (analgesic drug) and noscapine (anti-tussive drug). They are unique for the cell cycle and the process, because they block metaphase (M phase) in mitosis. They block tubulin's ability to form microtubules via polymerization. Dysfunctional spindle structures, avoid chromosome separation and cell proliferation. Strictosidine, which had been synthesized by novo, was the earliest study of alkaloid plant origin in *Saccharomyces cerevisiae* strains [14]. Strictosidine is a significant intermediate product in the terpenoids Indole alkaloids (TIAs) biosynthesis that include vincristine (antitumor) and ajmalicine (anti-hypertensive) [15].

The metabolic course of *Saccharomyces cerevisiae* for protoberberine alkaloid (S) canadine from racnorlaudanosolin has been optimized by Galanie et al. [16]. The secondary metabolite of Berberine has numerous pharmacological effects, such as antidiabetics, antibacterial, anti-ulcerones and anti-inflammatory effects. In vivo and vitro experiments with berberine, the results of arresting cell cycle during G1 or G2/M and apoptosis of tumor cell were shown to be anti-cancer activity. Berberine also was found to induce autophagy and stress in the endoplasmic reticulum, resulting in the invasion and inhibition of the tumor cells. Berberine has been proven to lower angiogenesis by decreasing expression of VEGF in addition to its apoptotic effects. The cancer cell migration was also decreased. In the respective

*DOI: http://dx.doi.org/10.5772/intechopen.95067*

complexes, Berberine attaches to DNA or RNA. Berberine also stimulates P53, which contributes to cell cycle arrest and apoptosis. Berberine has been shown to induce apoptosis as well via DNA interactions and pathways that are dependent on mitochondrial [17].

## 4. Phenolic

Phenolic are a large category of secondary plant metabolites that have at least one hydroxylated ring. Flavonoids and stilbenes are two groups of metabolites highly appreciated as a nutrient and therapeutic agent among the phenolic developed by phenylpropanoid pathways in the plant [18]. As a host cell, *Saccharomyces cerevisiae* is well equipped to promote phenolic compound biosynthesis. The following phases include cancer development: initiation, development, proliferation, invasion and metastasis. Initiation links by free radicals, inflammatory mediators, radiation, smoke of cigarette that damage the DNA-product, which can cause genetic mutation and replication of cells which are mutated results to cause carcinogenesis [19].

It has been noted that phenolic compounds i.e. ellagic acid and delphinidin show significant protective effects for example, apoptotic activity in colon cancer cells. In prostate, liver and leukemia cancer cells, delphinidin also showed significant activity. Cell apoptosis can also be caused by phenolic compounds. The mechanisms of invasion and migration of human cancer are also updated (**Figure 2**) [20].

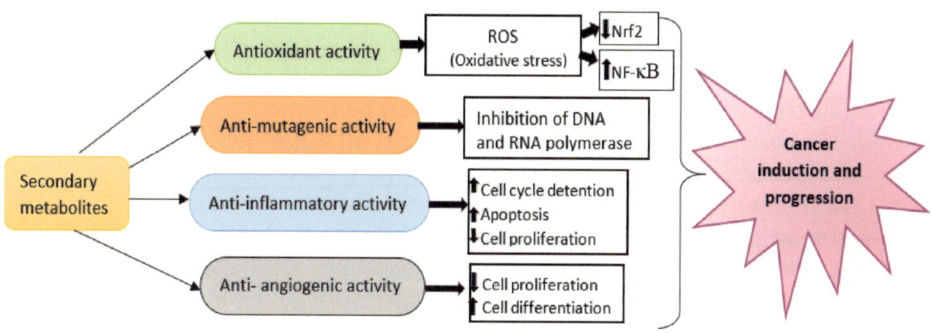


**Figure 2.**
*Potential of secondary metabolite protective mechanisms for cancer management [17].*

## 5. Flavonoids

The *Saccharomyces cerevisiae* strains have been developed by researchers, which synthesize various flavonoids, from the main phenolic intermediates to flavones, flavanones, flavonols and isoflavones. Phenylalanine is transformed into p-coumaric acid, a common phenylpropanoid precursor, in two sequence reactions in the initial steps of the phenylpropanoid pathway. *S. cerevisiae* is an ideal flavonoid host strain. One of the first studies on *S. cerevisiae* for flavonoid production was carried out according to [21]. The first flavonoid analysis was published on the synthesis of naringenin, major intermediate of flavonoid, from glucose in engineered *Saccharomyces cerevisiae* [22]. In *Saccharomyces cerevisiae* p-coumaric acid was used as a precursor for other flavonoids including apigenin, chrysin and luteolin. Anti-inflammatory, cardio-protective and anticancer effects were observed for quercetin, kaempferol, and fisetin [23].

More flavonoid intake has been reported to reduce cancer risk. In this respect, there have been reports of a variety of mechanisms including arresting of cell cycle, proliferation inhibition, apoptosis induction, anti-oxidation etc. These flavonoid extracts' cytotoxic activity makes them capable to produce cancer drugs. Flavonoids from *Saccharomyces cerevisiae* have potential to treat cancer treatment at all stages is therefore important for the recognition of harmless constituents against cancer as important for chemotherapy. Therefore, for producing cancer medicines, it is important to distinguish effective components from the yeast strain. Flavonoid-treated HeLa cells displayed apoptosis and loss of mitochondrial membrane potential (MMP). The toxic effect of flavonoid extracts makes them attractive candidates for cancer drug development. Such treatments have been found to inhibit growth across many lines of cancer cell, including cancer of the colon, breast cancer, carcinoma of squamous cell and hepatocellular carcinoma etc. Another study concerning bioactive flavonol "fisetin" found that fistein treatment caused cell viability by regulating the arresting G1 phase of human in melanoma cells [24].

## 6. Terpenoids

Many terpenoids were developed by engineering the related metabolic pathways in *Saccharomyces cerevisiae*. The first study about the medically useful development of terpenoids in yeast was the amorphadiene synthesis in *Saccharomyces cerevisiae*. The sigma ling of NF- κB, the key regulator in pathogenesis for cancer and inflammation, can be inhibited by natural terpenoids. Various mechanisms, such as the induction of apoptosis, have been found in terpenoid cancer. The natural terpenoids are quite well identified as NF-κB signaling inhibitors. Yeast also produces taxol (class of di-terpenoid), which is a common cancer medicine [25]. *Saccharomyces cerevisiae* has been metabolically adapted for the synthesis of taxa diene as a primary trial for taxol synthesis in a microbial host [26]. Paclitaxel (generally called taxol) is a common and strong drug used in chemotherapy for cancer. Taxol has the anti-tumor property due to its attachment to microtubule protein i.e. β-tubulin. As a result, the micro tubular dynamics have been suppressed and acetylation of α-tubulin protein has been increased. Mitosis can be prevented by increase in micro tubular and thus contributes to the cell's death.

The carotenoids were the first kind of isoprenoids formed in the *Saccharomyces cerevisiae*. These compounds are found to cardiovascular and osteoporosis antioxidant with therapeutic effects and to have anticancer activity by the activation of the NF-κB signaling pathway. LPS are reduced by signaling caused by β-carotene. It also reduces the IκB protein and prevents p65 subunit nuclear translocation and also prevent the NF-κB complex binding with DNA. β-carotene prevents cancer progression by virtue of its proxidant function [27].

## 7. Polyketides

Polyketides form a large group of bioactive compounds, which vary in pharmacology and structure, and cover different drugs like antibiotics and antitumor. The polyketides are synthesized by large enzymes called polyketide (PKS) as secondary metabolites in particular by fungi and bacteria. The Synthesis of dihydrochalcones (DHCs) is an example of the polyketide development of metabolic *Saccharomyces cerevisiae* strain. One polyketide derivative with antidiabetic and antioxidant activity is DHCs such as nothofagin and phlorizin. Preclinical trials indicate that its primary effect is to disrupt the dynamics of microtubule by reducing micro tubular polymerization and contribute to apoptosis. However, the mechanisms used to regulate eribulin's action were not fully elucidated [28]. Other pre-clinical studies in eribulin have demonstrated its ability in breast and colon cancer to suppress tumor growth in xenograft models of mouse. In metastatic breast cancer trials, clinical studies have shown that this compound is viable. Additional phase III analysis showed that eribulin as though only other two therapeutic drugs (taxane and anthracyclines) increases overall survival in pre-treated patients with severe breast cancer (**Figure 3**) [29].

Activation of the caspase is conducted along two separate routes: the cell membrane mediated pathway of the death recipient and the mitochondrial pathways. Rising Bax protein levels with a related decrease in the Bcl-2 protein have been shown by the two studies. A rise in the Bax/Bcl-2 ratio is well known to encourage cytochrome c release from mitochondria into the cytosol, culminating in the triggering of caspase-3, which is a cause for apoptosis initiation. A variety of signal transduction pathways and regulatory pathways contribute, among other things, to the upregulation of apoptosis genes are recorded to involve protein–protein interactions (PPIs) and their associated protein [30]. STAT-3 seem to have an effect on the migration of the tumor and glioblastoma cells, non-small cell lung cancer (NSCLC), renal system of human which have shown a metabolite disturb this

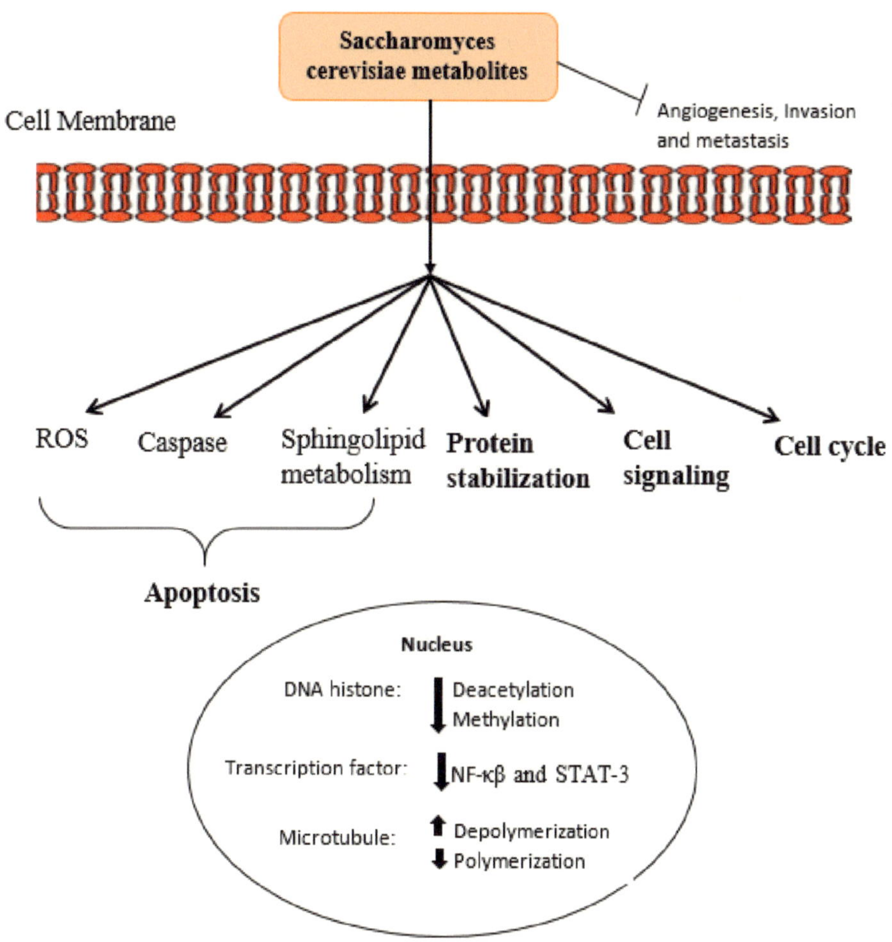


**Figure 3.**
*Targets of* Saccharomyces cerevisiae *secondary metabolites in tumor cells [9].*

system which indicates that the mechanism is likely to slow or suppress solid tumors in the metastatic process [31].

## 8. Non-ribosomal peptides (NRPSs)

Gurma is a 35-residue peptide widely used as a medical product in sweet-tasting transduction studies because of its ability to specifically inhibit neural responses for sweet matter in rats and mice is an example of the non-ribosomal peptide synthesis in modified yeast. Early research has shown that NRPSs are involved in the development of some of our key antiviral, anticancer, antibacterial and immunosuppressant medicines even though they are not progenitors of this ribosome. It has been determined the bleomycins are a group of glycosylated peptides which Umezawa and his colleagues found to be active against cancer in the 1960s. Bleomycin A2 and B2 is a major component of the clinical medicine, blenoxane. Carcinoma of cell squamous, lymphomas, esophageal cancer, testicular cancer and recalcitrant warts have been treated with blenoxane. This iron-bleomycin complex, mostly ferrous ions which interacts with oxygen and creates reactive free radical. Specifically, the free radicals respond to the abstraction of DNA deoxyribose hydrogen from C4' and lead to one or two-strand breaks, mainly of G-T and G-C at pyrimidines [32].

Several compounds for the production of novel anticancer drugs have recently come into being. However, the pathways involved in the cytotoxicity of these compounds in tumor cell lines are still widely ignored but many studies point to a significance in apoptosis. For example, a number of compounds impeding cell growth in a wide range of cell lines of cancer have been shown to be still poorly elucidated by which canner cells are hampered. Compounds were identified in some cases to cause cell death by the triggering of the apoptotic pathway; however, more studies were required in the mechanisms involved apoptosis. Some compounds created an imbalance in the ability of the cell redox, with mitochondria playing a key role in this phase. Further studies are required to explain this, however. A further impaired process is cell cycle, primarily as a result of actin and micro tubular filament interruption, yet only a few studies are conducted to link marine NRPs with cell cycle alterations, and further research are necessary to clarify their intervention in the process [33].

## 9. Vitamin C

L-ascorbic acid (Vitamin C and L-AA or) is a water-soluble effective antioxidant used as a ROS scavenger to defend against or at least reduce the harmful effects caused by ROS in most eukaryotic species. ROS is a common source of various types of cancer and considered vitamin C have a beneficial impact on reductions of lung, colorectal and stomach cancer [34]. On the other hand, vitamin C can, under some circumstances, enhance the production of ROS and may have adverse effects under some circumstances [35].

However, the ability of yeast cells to generate L-AA is normal. Erythro-ascorbic acid is instead present at a low concentration in yeast cells as a structurally connected substance with chemical composition identical to L-AA. The vitamin C biosynthesis from D-glucose in *S. cerevisiae* cells that was recombinant by using a plant paths is first recorded according to [36]. In two distinct strains, ascorbic acid accumulation was shown to be effective, and the effect from a distinct genetic context was studied in parallel [37].

## 10. Future perspectives

Further studies on the kinds of chemical components and the purification of different bioactive groupings could show the full capacity for certain pathogenic microbes inhibited by the *Saccharomyces cerevisiae* extract. The creation of a catalog of *Saccharomyces cerevisiae* processing hosts which are produced for the supply by the overproduction of central metabolites of sufficient precursors for heterologous pathways introduced can be aimed at future efforts in the sector. These modified strains often have the effect of modifying the functioning of organelle and cell membrane processes in order to enable the efficient use of substrates, product exports, intermediate retention and partitioning.

Additional crucial components are the compatibility between the introduced foreign secondary metabolite genes and the *Saccharomyces cerevisiae* host, including highly enhanced host pathway expression, metabolic stress/contamination, sufficient resistance/export of secondary metabolites etc. The strengthening of the hierarchical regulatory waterfalls within the host cell and the incorporation of appropriate promoters into the inserted BGC could help to improve the BGC expression for the secondary bioactive metabolite of an engineered host. Then, the resistant and exporting genes derived from or from original biosynthetic gene clusters or selected from other clusters of genes, other secondary source metabolites or even environment metagenomes might reduce the toxicity and metabolic burden on the host cell due to synthesis of secondary metabolites. In the coming years activities in other yeast species will be further investigated in order to identify additional results and obtain more high-performance products in this area instead of *Saccharomyces cerevisiae* in pharmaceutical secondary metabolites.

## 11. Conclusion

In summing up this paper, remarkable developments in the field of yeast cell and molecular biology have taken place in the last two decades, especially as cell engineering, genome sequencing and synthetic technology have grown rapidly. Many useful pharmaceuticals and metabolites were developed using modified *Saccharomyces cerevisiae*. These results support a strong prediction that secondary metabolites from *Saccharomyces cerevisiae* can be perceived to occur as single pharmacological compounds. The metabolites also target key regulatory pathways in cancer cells, normal cells and tissues. Until human trials are launched on cancer subjects, fungal metabolites must, however, undergo stringent quality control and pharmacological dose-scaling evaluations. In addition, high-end screening methods need to be standardized because new metabolites are constantly identified for the yeast metabolite library survey. Isolating and altering these secondary metabolites would enable the development of chemotherapy drugs.

## Author details

Muhammad Jahangeer[1], Areej Riasat[1], Zahed Mahmood[1], Muhammad Numan[1], Naveed Munir[1], Mehvish Ashiq[2], Muhammad Asad[1], Usman Ali[1] and Mahwish Salman[1*]

1 Department of Biochemistry, Government College University Faisalabad, Pakistan

2 Department of Chemistry, The Women University Multan, Pakistan

*Address all correspondence to: mahwish.gene@gmail.com

## References

[1] Hoffman CS, Wood V, Fantes PA. An ancient yeast for young geneticists: A primer on the Schizosaccharomyces pombe model system. Genetics. 2015;**201**(2):403-423

[2] Piškur J, Langkjaer RB. Yeast genome sequencing: The power of comparative genomics. Molecular Microbiology. 2004;*53*(2):381-389

[3] Ostergaard S, Olsson L, Nielsen J. Metabolic engineering of Saccharomyces cerevisiae. Microbiology and Molecular Biology Reviews. 2000;*64*(1):34-50

[4] Chin YW, Balunas MJ, Chai HB, Kinghorn AD. Drug discovery from natural sources. The AAPS Journal. 2006;*8*(2):E239-E253

[5] Al-Jassaci MJ, Mohammed GJ, Hameed IH. Secondary metabolites analysis of saccharomyces cerievisiae and evaluation of antibacterial activity. International Journal of Pharmaceutical and Clinical Research. 2016;*8*(5):304-315

[6] Park SR, Yoon YJ, Pham JV, Yilma MA, Feliz A, Majid MT, et al. A review of the microbial production of bioactive natural products and biologics. Frontiers in Microbiology. 2019;*10*:1404

[7] Fidan O, Zhan J. Recent advances in engineering yeast for pharmaceutical protein production. RSC Advances. 2015;**5**(105):86665-86674

[8] Thomas DL, Sharp TM, Torres J, Armstrong PA, Munoz-Jordan J, Ryff KR, et al. Local transmission of zika virus—puerto rico. Morbidity and Mortality Weekly Report. November 23, 2015–January 28, 2016. 2016;65(6):154-158

[9] Banerjee S, Paruthy SB. Preclinical and clinical perspective on fungal metabolites and their analogs as anticancer agents–from bench to bedside. Fungal Metabolites. 2016:1-32

[10] Rabha J, Jha DK. Metabolic diversity of penicillium. In: *New and Future Developments in Microbial Biotechnology and Bioengineering*. Elsevier; 2018. pp. 217-234

[11] Rahmat E, Kang Y. Yeast metabolic engineering for the production of pharmaceutically important secondary metabolites. Applied Microbiology Biotechnology. 2020;**104**(11):4659-4674. https://doi.org/10.1007/s00253-020-10587-y

[12] Rahmat E, Kang Y. Yeast metabolic engineering for the production of pharmaceutically important secondary metabolites. Applied Microbiology and Biotechnology. 2020:1-16

[13] Siddiqui MS, Thodey K, Trenchard I, Smolke CD. Advancing secondary metabolite biosynthesis in yeast with synthetic biology tools. FEMS Yeast Research. 2012;**12**(2):144-170

[14] Brown S, Clastre M, Courdavault V, O'Connor SE. De novo production of the plant-derived alkaloid strictosidine in yeast. Proceedings of the National Academy of Sciences. 2015;**112**(11):3205-3210

[15] Nam KH, Chung HJ, Jeon EJ, Park MK, Yim YH, Liu JR, et al. In vitro biosynthesis of strictosidine usinglonicera japonica leaf extracts and recombinant yeast. Journal of Plant Biology. 2007;**50**(3):315-320

[16] Galanie S, Thodey K, Trenchard IJ, Interrante MF, Smolke CD. Complete biosynthesis of opioids in yeast. Science. 2015;**349**(6252):1095-1100

[17] Kaboli PJ, Rahmat A, Ismail P, Ling KH. Targets and mechanisms of

berberine, a natural drug with potential to treat cancer with special focus on breast cancer. European Journal of Pharmacology. 2014;**740**:584-595

[18] Ververidis F, Trantas E, Douglas C, Vollmer G, Kretzschmar G, Panopoulos N. Biotechnology of flavonoids and other phenylpropanoid-derived natural products. Part I: Chemical diversity, impacts on plant biology and human health. Biotechnology journal: Healthcare nutrition. Technology. 2007;**2**(10):1214-1234

[19] Vejarano R, Gil-Calderón A, Díaz-Silva V, León-Vargas J. Improvement of the bioactive profile in wines and its incidence on human health: Technological strategies. In: Advances in Grape and Wine Biotechnology. IntechOpen; 2019

[20] Prasad R, Katiyar SK. Grape seed proanthocyanidins inhibit migration potential of pancreatic cancer cells by promoting mesenchymal-to-epithelial transition and targeting NF-κB. Cancer Letters. 2013;**334**(1):118-126

[21] Coates AS, Winer EP, Goldhirsch A, Gelber RD, Gnant M, Piccart-Gebhart M, et al. Tailoring therapies—improving the management of early breast cancer: St Gallen International Expert Consensus on the Primary Therapy of Early Breast Cancer 2015. Annals of oncology. 2015;**26**(8):1533-1546. https://doi.org/10.1093/annonc/mdv221

[22] Koopman, F., Beekwilder, J., Crimi, B., van Houwelingen, A., Hall, R. D., Bosch, D., ... & Daran, J. M. (2012). De novo production of the flavonoid naringenin in engineered Saccharomyces cerevisiae. Microbial Cell Factories, 11(1), 155.

[23] Ye X, Zheng C, Chen J, Gao Y, Murray CB. Using binary surfactant mixtures to simultaneously improve the dimensional tunability and monodispersity in the seeded growth of gold nanorods. Nano Letters. 2013;**13**(2): 765-771. https://doi.org/10.1021/nl304478h

[24] Hayat M, Abbas M, Munir F, Hayat MQ, Keyani R, Amir R. Potential of plant flavonoids in pharmaceutics and nutraceutics. Journal of Biomolecules and Biochemistry. 2017;**1**(1):12-17

[25] Ye VM, Bhatia SK. Metabolic engineering for the production of clinically important molecules: Omega-3 fatty acids, artemisinin, and taxol. Biotechnology Journal. 2012;**7**(1):20-33

[26] Engels B, Dahm P, Jennewein S. Metabolic engineering of taxadiene biosynthesis in yeast as a first step towards Taxol (paclitaxel) production. Metabolic Engineering. 2008;**10**(3-4): 201-206

[27] Jain H, Dhingra N, Narsinghani T, Sharma R. Insights into the mechanism of natural terpenoids as NF-κB inhibitors: An overview on their anticancer potential. Experimental Oncology. 2016

[28] Smith JA, Wilson L, Azarenko O, Zhu X, Lewis BM, Littlefield BA, et al. Eribulin binds at microtubule ends to a single site on tubulin to suppress dynamic instability. Biochemistry. 2010; **49**(6):1331-1337

[29] Lorente A, Makowski K, Albericio F, Álvarez M. Bioactive marine polyketides as potential and promising drugs. Annals of Marine Biology and Research. 2014;**1**(1):1-10

[30] Shrestha G, Clair LLS. Lichens: A promising source of antibiotic and anticancer drugs. Phytochemistry Reviews. 2013;**12**(1):229-244

[31] Ierano C, Basseville A, To, K. K, Zhan Z, Robey RW, Wilkerson J, et al. Histone deacetylase inhibitors induce CXCR4 mRNA but antagonize CXCR4

migration. Cancer Biology & Therapy. 2013;**14**(2):175-183

[32] Felnagle EA, Jackson EE, Chan YA, Podevels AM, Berti AD, McMahon MD, et al. Nonribosomal peptide synthetases involved in the production of medically relevant natural products. Molecular Pharmaceutics. 2008;**5**(2):191-211

[33] Agrawal S, Acharya D, Adholeya A, Barrow CJ, Deshmukh SK. Nonribosomal peptides from marine microbes and their antimicrobial and anticancer potential. Frontiers in Pharmacology. 2017;**8**:828

[34] Valko M, Rhodes C, Moncol J, Izakovic MM, Mazur M. Free radicals, metals and antioxidants in oxidative stress-induced cancer. Chemico-Biological Interactions. 2006;**160**(1):1-40

[35] Naidu KA. Vitamin C in human health and disease is still a mystery? An overview. Nutrition Journal. 2003;**2**(1):7

[36] Branduardi P, Fossati T, Sauer M, Pagani R, Mattanovich D, Porro D. Biosynthesis of vitamin C by yeast leads to increased stress resistance. PLoS One. 2007;**2**(10):e1092

[37] Huang B, Guo J, Yi B, Yu X, Sun L, Chen W. Heterologous production of secondary metabolites as pharmaceuticals in Saccharomyces cerevisiae. Biotechnology Letters. 2008;**30**(7): 1121-1137